P9-CJY-534

A GUIDE TO HAZARDOUS MATERIALS MANAGEMENT

Recent Titles from Quorum Books

Training in the Automated Office: A Decision-Maker's Guide to Systems Planning and Implementation
Randy J. Goldfield

The Strategist CEO: How Visionary Executives Build Organizations
Michel Robert

America's Future in Toxic Waste Management: Lessons from Europe
Bruce W. Piasecki and Gary A. Davis

Telecommunications Management for the Data Processing Executive: A Decision-Maker's Guide to Systems Planning and Implementation
Milburn D. Smith III

The Political Economy of International Debt: What, Who, How Much, and Why?
Michel Henri Bouchet

Multinational Risk Assessment and Management: Strategies for Investment and Marketing Decisions
Wenlee Ting

Supply-Side Portfolio Strategies
Victor A. Canto and Arthur B. Laffer, editors

Staffing the Contemporary Organization: A Guide to Planning, Recruiting, and Selecting for Human Resource Professionals
Donald L. Caruth, Robert M. Noe III, and R. Wayne Mondy

The Court and Free-Lance Reporter Profession: Improved Management Strategies
David J. Saari

Pollution Law Handbook: A Guide to Federal Environmental Laws
Sidney M. Wolf

Res Judicata and Collateral Estoppel: Tools for Plaintiffs and Defendants
Warren Freedman

The Investment Performance of Corporate Pension Plans: Why They Do Not Beat the Market Regularly
Stephen A. Berkowitz, Louis D. Finney, and Dennis E. Logue

Foreign Plaintiffs in Products Liability Actions: The Defense of Forum Non Conveniens
Warren Freedman

A GUIDE TO HAZARDOUS MATERIALS MANAGEMENT

Physical Characteristics, Federal Regulations, and Response Alternatives

AILEEN SCHUMACHER

QUORUM BOOKS
New York • Westport, Connecticut • London

Library of Congress Cataloging-in-Publication Data

Schumacher, Aileen.
A guide to hazardous materials management.

Bibliography: p.
Includes index.
1. Hazardous substances—United States—Management.
2. Hazardous wastes—United States—Management.
I. Title.
TD811.5.S35 1988 363.1'7 87-7227
ISBN 0-89930-255-6 (lib. bdg. : alk. paper)

British Library Cataloguing in Publication Data is available.

Library of Congress Catalog Card Number: 87-7227
ISBN: 0-89930-255-6

First published in 1988 by Quorum Books

Greenwood Press, Inc.
88 Post Road West, Westport, Connecticut 06881

Printed in the United States of America

The paper used in this book complies with the Permanent Paper Standard issued by the National Information Standards Organization (Z39.48-1984).

10 9 8 7 6 5 4 3 2 1

Copyright Acknowledgments

The following publishers have generously granted permission to use material from copyrighted work:

Table 2.3, "Radiation Exposure Limitations," Figure 11.4, "Appropriate Fire Extinguishers for Various Types of Fires," and Table A.16, "Some Incompatible Chemicals," from *Safety Engineering Standards*, copyright © 1980 by Industrial Indemnity Company. Reprinted by permission of the publisher.

Table 10.1, "Hazardous Wastes from Everyday Products," from *Hazardous Waste Management—In Whose Backyard?* copyright © 1984 by Westview Press. Reprinted by permission of the publisher.

Table 10.6, "Household Hazardous Waste Chart," from *Hazardous Waste What You Should & Shouldn't Do* (a brochure), copyright © 1986 by the Water Pollution Control Federation. Reprinted by permission of the publisher.

Table 11.4, "Protective Clothing Materials Rated by Chemical," from *Leachate from Hazardous Wastes Sites,* copyright © 1983 by Technomic Publishing Company. Reprinted by permission of the publisher.

Figure 11.2, "NFPA 704 Marking System," adapted with permission from material with copyright © 1985, National Fire Protection Association.

Extended quotations appear in Chapter 10 from the following works:

Engineering Times, Vol. 6, No. 4 (April 1984), p. 3, copyright © 1984 by the National Society of Professional Engineers. Reprinted by permission of the editor.

Siting Hazardous Waste Management Facilities, copyright © 1983 by the Conservation Foundation. Reprinted by permission of the publishers: the Conservation Foundation, Chemical Manufacturers Association, and National Audubon Society.

The book is dedicated to Nicky and Kevin, with the fervent hope that they will be afforded the chance to grow up in a safe and pleasing environment. They have certainly brightened mine.

FAMOUS QUOTATIONS

On regulatory policy:

The sun too penetrates into privies, but it is not polluted by them.

—Diogenes The Cynic

On human error in chemical analyses:

Elmer was a chemist
But Elmer is no more
For what Elmer thought
Was H_2O was H_2SO_4

—Anonymous

On preserving the environment for future generations:

What has our progeny ever done for us?

—Anonymous

On landfill inspections:

I love all waste and solitary places . . .

—Shelley

On the need for good public relations:

What kills a skunk is the publicity it gives itself.

—Abraham Lincoln

On the relationship of politics and the environment:

Avoid the Reeking Herd
Shun the Polluted Flock . . .

—Elinor Hoyt Wylie

On the warnings of inspectors:

No lie you can speak or act but it will come, after longer or shorter circulation, like a bill drawn on Nature's Reality, and be presented there for payment . . .

—Thomas Carlyle

CONTENTS

Figures and Tables xiii
Acknowledgments xvii
Acronyms xix
1. Introduction 1
2. What Makes a Material Hazardous? 5
3. The Rules of the Game: Regulations 19
4. Historical Precedent Regulations 29
5. Air and Water Quality Regulations 49
6. Toxic/Hazardous Materials Regulations 77
7. Asbestos Control Regulations 109
8. Radioactive Materials Control Regulations 117
9. Program Management 125
10. Substance Management 135
11. Emergency Response 157
12. Regulatory Compliance and Work Practices 171
13. Sources of Additional Information 177
Appendix: Supplementary Technical Information 181
Glossary 267
Selected Bibliography 277
Index 283

FIGURES AND TABLES

FIGURES

2.1 Flowchart of the Sequence of Events During a Controlled Series of Laboratory Measurements 16
2.2 Accuracy and Precision 18
3.1 The Process of Rule Making 21
3.2 Hazardous Materials in the Regulatory Framework 24
4.1 Material Safety Data Sheet 40
5.1 Standards Setting Procedure for NAAQS 51
5.2 Standards Setting Procedure for NESHAPs 56
5.3 Standards Setting Procedure for NSPS 59
5.4 State PSD Program Status 61
5.5 Relationship of Elements Used in Defining NPDES Permit Conditions 66
6.1 DOT Specified 17H Steel Drum (30 Gallon) 79
6.2 DOT Specified 7A Steel-Banded Wooden Shipping Container 80
6.3 Uniform Hazardous Waste Manifest 85
6.4 Identification of a RCRA Solid Waste 94
6.5 Identification of a RCRA Hazardous Waste 95
6.6 Superfund Response 103
7.1 Initial Steps in an ACM Survey 112
10.1 Hazardous Materials Management Technologies 137

11.1 National Contingency Plan Operation 159
11.2 NFPA 704 Marking System 161
11.3 Simplified Hazardous Materials Spill Response Diagram 165
11.4 Appropriate Fire Extinguishers for Various Types of Fires 167

TABLES

2.1 Some Common Water Reactive Materials 8
2.2 Some Common Oxidizing Agents 9
2.3 Radiation Exposure Limitations 14
3.1 Imminent Hazard Provisions of Federal Statutes 23
3.2 Federal Laws Addressing Hazardous Materials Management 25
3.3 Other Laws Affecting Hazardous Materials 26
4.1 Common Pesticides 31
4.2 Leading Work-Related Diseases and Injuries 35
4.3 Status of State Plan States' Response to Hazard Communication Standard 42
5.1 Criteria Pollutants Regulated by NAAQS 52
5.2 Pollutants and Sources Regulated under NESHAPs 55
5.3 Summary of CAA Motor Vehicle Regulations 57
5.4 Pollutants Regulated under the CAA 60
5.5 CWA State Program Status 68
5.6 Primary Drinking Water Standards 71
5.7 SDWA Well Classifications 73
5.8 UIC State Programs 75
6.1 DOT Hazard Classes 81
6.2 Comparison of EPA's Hazardous Waste Characteristics with DOT's Hazard Classes 82
6.3 Descriptive Summary of DOT Labels 83
6.4 Generator and Transporter Standards 97
6.5 General TSDF Requirements 98
6.6 Types of Response Actions 102
9.1 Selected Hazardous Materials Management Programs 132
10.1 Hazardous Wastes from Everyday Products 136

10.2	Proper Procedures for Storage of Hazardous Materials	138
10.3	Stabilization/Solidification Processes	144
10.4	Common Solid Waste Incinerator Types	147
10.5	Selected Innovative Treatment/Clean-up Technologies	149
10.6	Household Hazardous Waste Chart	154
11.1	Federal Agencies and Departments Governing Hazardous Materials	158
11.2	UN Hazard Class Numbers	160
11.3	Sample Telephone Roster for Emergency Responses	163
11.4	Protective Clothing Materials Rated by Chemical	168
A.1	Priority Pollutants	182
A.2	Reportable Quantities (RQs)	185
A.3	Clean Water Act (1987) Funding	189
A.4	Drinking Water Contaminants for which Standards Must Be Established by EPA	191
A.5	DOT Hazard Classifications	193
A.6	Hazardous Wastes from Specific Sources	197
A.7	Acute Hazardous Wastes	205
A.8	Hazardous Wastes from Nonspecific Sources	209
A.9	Toxic Hazardous Wastes	213
A.10	National Priorities List: Non-Federal Sites	222
A.11	National Priorities List: Federal Sites	240
A.12	CERCLA Extremely Hazardous Substances	241
A.13	Hazardous Substances Prioritized for Further Study	253
A.14	Asbestos-Containing Materials Found in Buildings	255
A.15	Asbestos Abatement Methods for Surfacing Materials	258
A.16	Some Incompatible Chemicals	261

ACKNOWLEDGMENTS

I would like to acknowledge the University of Florida, which conducted the training session that this book was originally conceived to supplement. In conjunction with those initial efforts, special thanks go to Stephanie Simmons-West and Evelyn Knauft.

This publication would not have been possible without the dedicated effort of members of my staff: Judy McNeill, Ken Sunseri, Sara Gibson, and Bill TenBroeck. I also gratefully acknowledge the help, support, and comments of Richard R. Blum, P.E.

I want to express my appreciation to those outstanding professionals of my acquaintance who reviewed and thereby improved portions of this book:

—Chris Bird
Environmental Engineer
Department of Environmental Services
Alachua County, Florida

—David A. Buff, M.E., P.E.
Principal Engineer
KBN Engineering and Applied Sciences, Inc.

—Leonard Carter, P.E.
Staff Engineer
Environmental Science & Engineering, Inc.

—Charlie Dougherty, P.E.
Environmental Engineer
Montgomery Engineers, Inc.

—Ernie Frey
District Manager
Florida Department of Environmental Regulation

—Tom Park
Environmental Chemist
PPB Environmental Laboratories, Inc.

—Bernhart C. Warren
Health Physicist
Quadrex HPS, Inc.

Special thanks in this regard also go to Trish Markey, P.E.

Finally, I want to acknowledge the part of Dr. James E. Steelman, who taught me my first course in engineering and thereby started it all.

ACRONYMS

Nature is . . . the mere supplier of that cunning alphabet, whereby selecting and combining as he pleases, each man reads his own peculiar mind and mood.

—Herman Melville

ACM	Asbestos Containing Materials
AEA	Atomic Energy Act
AEC	Atomic Energy Commission
AHERA	Asbestos Hazard Emergency Response Act
ANPR	Advance Notice of Proposed Rulemaking
APA	Administrative Procedure Act
AQCR	Air Quality Control Region
ASHAA	Asbestos School Hazard Abatement Act
ASME	American Society of Mechanical Engineers
BACT	Best Available Control Technology
BADCT	Best Available Demonstrated Control Technology
BAT	Best Available Technology
BCT	Best Conventional Technology
BEJ	Best Engineering Judgment
BLEVE	Boiling Liquid Expanding Vapor Explosion
BMP	Best Management Practice(s)
BPT	Best Practicable Technology
BRH	Bureau of Radiological Health
CAA	Clean Air Act
CAS	Chemical Abstract Services

CBI	Confidential Business Information
CEQ	Council on Environmental Quality
CERCLA	Comprehensive Environmental Response, Compensation, and Liability Act
CFC	Chlorofluorocarbon
CFR	*Code of Federal Regulations*
CHEMTREC	Chemical Transportation Emergency Center
CHLOREP	Chlorine Emergency Plan
CHLW	Commercial High Level Waste
CHRIS	Chemical Hazards Response Information System
CICIS	Chemicals in Commerce Information System
CMA	Chemical Manufacturers Association
COE	U.S. Army Corps of Engineers
CPSC	Consumer Product Safety Commission
CSIN	Chemical Substances Information Network
CWA	Clean Water Act
DCE	1,1-Dichloroethylene
DHHS	Department of Health and Human Services
DHLW	Defense High Level Waste
DOC	Department of Commerce
DOD	Department of Defense
DOE	Department of Energy
DOJ	Department of Justice
DOL	Department of Labor
DOS	Department of State
DOT	Department of Transportation
EDF	Environmental Defense Fund
EIS	Environmental Impact Statement
EP	Extraction Procedure
EPA	Environmental Protection Agency
EPCRA	Emergency Planning and Community Right-to-Know Act
ERDA	Energy Research and Development Administration
FAA	Federal Aviation Administration
FDA	Food and Drug Administration
FEMA	Federal Emergency Management Agency
FEPCA	Federal Environmental Pesticide Control Act
FFDCA	Federal Food, Drug, and Cosmetic Act
FHWA	Federal Highway Administration

FIFRA	Federal Insecticide, Fungicide, and Rodenticide Act
FR	*Federal Register*
FRA	Federal Railway Administration
FRC	Federal Radiation Council
FS	Feasibility Study
FWPCA	Federal Water Pollution Control Act
FWS	Fish and Wildlife Service
HCS	Hazard Communication Standard
HELP	Hazardous Emergency Leak Procedure
HEW	Health, Education, and Welfare
HHS	Health and Human Services
HLW	High Level Waste
HMTA	Hazardous Materials Transportation Act
HMTR	Hazardous Materials Transportation Regulations
HRCQ	Highway Route Controlled Quantities
HRS	Hazard Ranking System
HSWA	Hazardous and Solid Waste Amendments
IAEA	International Atomic Energy Agency
IDLH	Immediately Dangerous to Life or Health
IOC	Inorganic Chemicals
IRAP	Interagency Radiological Assistance Plan
ITSDC	Interagency Toxic Substances Data Committee
LAER	Lowest Achievable Emission Rate
LC50	Lethal Concentration, 50 percent kill
LD50	Lethal Dose, 50 percent kill
LEL	Lower Explosive Limit
LIP	Label Improvement Program
LLW	Low Level Waste
LSA	Low Specific Activity
LUST	Leaking Underground Storage Tank
MCL	Maximum Contaminant Level
MCLG	Maximum Contaminant Level Goal
MPRSA	Marine Protection, Research, and Sanctuaries Act
MRS	Monitored Retrievable Storage
MSDS	Material Safety Data Sheet
MSHA	Mine Safety and Health Administration
MTB	Materials Transportation Bureau
NA	Nonattainment Area

NA	North American (identification number)
NAAQS	National Ambient Air Quality Standards
NCP	National Contingency Plan
NEPA	National Enviromental Policy Act
NESHAP	National Emission Standards for Hazardous Air Pollutants
NFPA	National Fire Protection Association
NIMBY	Not In My Backyard!
NIOSH	National Institute for Occupational Safety and Health
NOAA	National Oceanic and Atmospheric Administration
NOS	Not Otherwise Specified
NPDES	National Pollutant Discharge Elimination System
NPL	National Priorities List
NRC	National Response Center
NRC	Nuclear Regulatory Commission
NRT	National Response Team
NSPS	New Source Performance Standards
OFA	Office of Federal Activities
OHMR	Office of Hazardous Materials Regulations
OHM-TADS	Oil and Hazardous Materials Technical Assistance Data System
OOE	Office of Operations and Enforcement
OPP	Office of Pesticide Programs
OPTS	Office of Pesticides and Toxic Substances
ORM	Other Regulated Material
OSC	On-Scene Coordinator
OSHA	Occupational Safety and Health Administration
OSH Act	Occupational Safety and Health Act
OTA	Office of Technology Assessment
PA	Preliminary Assessment
PBB	Polybrominated Biphenyl
PCB	Polychlorinated Biphenyl
PEL	Permissible Exposure Limit
PMN	Premanufacturing Notice
POHC	Primary Organic Hazardous Constituent
POTW	Publicly Owned Treatment Works
PPM	Parts per Million
PSD	Prevention of Significant Deterioration
PSTN	Pesticide Safety Team Network
QA	Quality Assurance

QC	Quality Control
RACT	Reasonably Available Control Technology
RBE	Relative Biological Effectiveness
RCRA	Resource Conservation and Recovery Act
REL	Recommended Exposure Limit
RI	Remedial Investigation
RMCL	Recommended Maximum Contaminant Level
ROD	Record of Decision
RPAR	Rebuttal Presumption against Registration
RQ	Reportable Quantity
RRC	Regional Response Center
RRT	Regional Response Team
RSPA	Research and Special Programs Administration
SARA	Superfund Amendments and Reauthorization Act
SDWA	Safe Drinking Water Act
SI	Site Inspection
SIC	Standard Industrial Classification
SIP	State Implementation Plan
SNARL	Suggested No Adverse Response Level
SNUR	Significant New Use Regulation
SOC	Synthetic Organic Chemicals
SPCC	Spill Prevention Control and Countermeasure
SQG	Small Quantity Generator
STEL	Short Term Exposure Limit
SWDA	Solid Waste Disposal Act
TCA	1,1,1-Trichloroethane
TCE	Trichloroethylene
TERP	Transportation Emergency Reporting Procedure
TLV	Threshold Limit Value
TPQ	Threshold Planning Quantity
TRU	Transuranic Waste
TSCA	Toxic Substances Control Act
TSDF	Treatment, Storage, and/or Disposal Facility
TWA	Time Weighted Average
UEL	Upper Explosive Limit
UIC	Underground Injection Control
UN	United Nations (identification number)
USC	United States Code

USCG	U.S. Coast Guard
USDA	U.S. Department of Agriculture
USGS	U.S. Geological Survey
USN	U.S. Navy
UST	Underground Storage Tank
VOC	Volatile Organic Compound
WQA	Water Quality Act
WPCF	Water Pollution Control Federation

A GUIDE TO HAZARDOUS MATERIALS MANAGEMENT

1

INTRODUCTION

> Hazardous materials management is the application of engineering and managerial control techniques to identify, evaluate and eliminate or reduce risks, involving conditions and practices related to hazardous materials . . .
>
> —Board of Hazard Control Management, 1986

Public awareness about hazardous materials has increased dramatically since 1970, when the annual report by the President's Council on Environmental Quality did not contain a single reference to hazardous waste. The purpose of this book is to explain the fundamental principles involved in hazardous materials management. It is written for a diversity of readers, including the citizen who wants to better understand the issues behind the publicity, the manager who is suddenly faced with employee safety considerations, the executive concerned with his industry's environmental profile, and the professional who is an expert in one aspect of this field, but desires a broader knowledge of other aspects of the subject.

In order to fully understand any subject, it is imperative to be able to distinguish between and correctly utilize the terms involved. ''Hazard'' is, in itself, a rather general term, subject to a great deal of interpretation. Material properties that result in hazards are discussed in Chapter 2, to better define exactly what it is that makes something hazardous.

When the adjective ''hazardous'' is coupled with ''material,'' ''substance,'' or ''waste,'' it becomes a term with specific legal meaning. *Hazardous material* is a broad term encompassing any material, including substances and wastes, that may pose an unreasonable risk to health, safety, property, or the environment when they exist in specific quantities and forms. *Hazardous substances* are a subset of hazardous materials, which are identified and regulated by the U.S. Environmental Protection Agency

(EPA) under various environmental regulations. *Hazardous wastes* are discarded materials that pose a risk to human health, safety, property, or the environment.

Toxicity is another term commonly used in discussions of hazardous materials, and it refers to the capacity to cause toxic effects in the living organisms, including humans. *Toxic materials* are all chemical substances and mixtures, including wastes, which have toxic properties. *Toxic pollutants* are a specific list of 65 chemicals or chemical classes identified under the Clean Water Act (CWA) and regulated by EPA. When broken out into resulting individual chemicals, these pollutants are known as priority pollutants. *Toxic substances* are chemicals that are subject to regulations issued under the Toxic Substances Control Act (TSCA), which is administered by EPA. This term may also be used in a generic sense to mean "toxic chemicals" or "toxic agents." *Toxic wastes* are those hazardous wastes that are listed as toxic or meet the characteristics of toxicity as defined in the Resource Conservation and Recovery Act (RCRA).

These terms are fundamental to an understanding of the "rules of the game," which are the environmental regulations governing the lawful management of hazardous materials. Chapter 3 discusses the fundamentals of such rule making, and Chapters 4 through 8 summarize selected major federal laws dealing with hazardous materials.

It should be emphasized that, like all human institutions, this body of laws is by no means an absolute. The law of gravitation and Newton's three laws of motion do not change, nor can they be negotiated. Human laws do not share these characteristics. Environmental laws are dynamic, subject to change, interpretation, and negotiation. Although an overview of regulations presents valid information, exceptions are born of generalities, and whether dealing with this book or an issue of the *Federal Register* (FR), it is imperative to make sure that judgments are based on the current version of the law. The reader is advised to always check to determine if any pertinent updates or revisions are in effect.

Management, the selection of best alternatives, is discussed in Chapters 9 and 10. Everyone needs to know what to do when things go wrong or, better yet, what to do before things go wrong. This subject is addressed in Chapter 11. Chapter 12 discusses compliance and best work practices. These two topics are nearly impossible to address comprehensively, as their definition normally depends on facility-specific conditions. However, basic concepts have been summarized and presented with a discussion of potential approaches constituting best work practices. Chapter 13 lists useful information sources. Technical information of a detailed or lengthy nature referenced in the text has been compiled in the appendix, which is followed by a glossary.

A word about acronyms: Their use is ubiquitous, extensive, often confusing, and almost unavoidable in any discussion of hazardous materials management. An acronym is a two-sided tool. On the positive side, a working knowledge of acronyms can help to decipher or express information in a very concise manner. However, an assumed knowledge of acronyms can baffle readers who would know the answer to their questions if only they knew what the letters stood for. And incorrect usage can baffle everyone, whether in written text or in conversation: Note the similarity in the pronunciations of the acronyms for the Occupational Safety and Health Administration and the Asbestos School Hazard Abatement Act of 1984 (OSHA and ASHAA).

To aid the reader who may wish to use only portions of this book, the following practice regarding acronyms has been adopted: Complete terms will be spelled out the first time they are used in every chapter or in every subchapter devoted entirely to a discussion of a specific regulation. The one exception to this rule is the use of EPA to represent Environmental Protection Agency throughout the entire text. For convenient reference, a comprehensive list of pertinent acronyms appears in the front matter of this book.

As with any dynamic subject matter, some of the information presented here will require updating as time progresses. The reader should take care to verify the status of specific regulations and any related judicial interpretations at the time of consideration. For this purpose, consultation of the *Code of Federal Regulations* (CFR) and FR is invaluable. The CFR contains the regulations promulgated under statute and is compiled annually every July. The FR contains proposed changes and updates to the regulations, in addition to background and explanatory passages. (More information on accessing regulations is presented in Chapter 3.)

Portions of environmental statutes are codified and incorporated into the CFR in whole or in part when their content is judged to be, in fact, regulatory. Entire statutes can be obtained from the U.S. Government Printing Office by reference to their Public Law number. Sometimes very recent statutes can be obtained from the offices of congressional officials before they become otherwise available. For statutes which are not very recent, it is advisable to consult the United States Code (USC), which will contain amendments made to the law after it was passed.

In addition to these printed sources of information, EPA staffs some toll-free telephone numbers. These, along with telephone numbers for other informational sources, are compiled in Chapter 13 and listed by subject in the text.

2

WHAT MAKES A MATERIAL HAZARDOUS?

Science is the first word on everything, and the last word on nothing.
—Victor Hugo

INTRODUCTION

Before beginning any discussion about the properties of hazardous materials, it is desirable to review the relative nature of the term "hazardous." A general but workable definition of hazard is "a condition that can be expected to cause damage(s), including injury or death, to exposed individuals." To determine if a material constitutes such a condition, due consideration must be given to its quantity and form.

Consider the following example: A student sits reading a chemistry text in a college laboratory. On the table next to him sits a sealed vial containing a gram of mercury. This mercury poses no substantial risk to the student.

But later, as a prank, someone breaks into the laboratory and steals the vial of mercury. This person pours it into the reservoir that supplies water to the dormitory where the student resides. The water chemistry is such that the mercury dissolves in the water. The student drinks enough of the water that he begins to experience the symptoms of mercury poisoning. The same substance has, in this instance, become a very hazardous material.

This example involved a commonly known, familiar material. Safety considerations in other situations are complicated by the fact that the number of chemical substances we can synthesize from natural products is virtually unlimited, and new products are continually being created. It is possible, however, to consider a finite number of broad categories that describe the manner in which a chemical or chemical mixture may be hazardous:

1. The material presents a fire danger. It is explosive, flammable, or undergoes a reaction which releases heat.

2. The material is a cryogen. It presents a hazard because of its extremely cold physical state.
3. The material is toxic. Depending on the concentration, duration, and type of exposure, the material may cause sickness or even death.
4. The material is radioactive. Exposure induces reactions in human tissues which can cause cell death. (The effects of radioactivity can also be considered a special type of toxicity.)

It has been suggested in recent years that a new hazardous category be added: stress. Using a working definition of stress as the rate of "wear and tear" on the body, it appears reasonable that in certain situations working with hazardous materials may exert damage which is independent of the categories just described. Therefore, these five categories of properties of hazardous materials are discussed in this chapter, in addition to some analytical considerations related to identifying hazardous materials.

CATEGORIES OF HAZARD

Fire Dangers

An understanding of numerous fire hazards may be gained from an understanding of the fundamentals of chemistry and physics which describe the properties of matter. Although the phenomenon of burning materials is a familiar one, it is not always obvious to the casual observer that, generally, liquids and solids do not actually burn as the liquid or solid state of matter. Instead, they give off vapors that ignite when a combustible mixture with air has been obtained. Many fire hazard definitions are based on vapor ignition, and many of these terms are used to describe the degree of danger associated with flammable materials:

—*Lower Explosive Limit (LEL)*: the minimum concentration of gas or vapor in air below which a substance does not burn when exposed to an ignition source.

—*Upper Explosive Limit (UEL)*: maximum concentration of gas or vapor in air above which ignition does not occur.

—*Flammable range*: the numerical difference between the upper and lower explosive limits.

—*Flash point*: the minimum liquid temperature at which a spark or flame will cause an instantaneous flash in the vapor space above the liquid.

—*Kindling point*: minimum solid temperature at which sufficient vapors exist to burn with a flame.

—*Fire point*: the temperature at which the liquid gives off enough vapor to continue to burn when lighted.

—*Spontaneous combustion*: the minimum temperature at which ignition occurs without the introduction of an ignition source (synonymous with auto ignition temperature).

—*Pyrophoric*: substances that spontaneously ignite in air without an ignition source.

Many hazardous materials have fire danger-associated properties. These include water reactive materials, oxidizing agents, organic compounds, polymers, corrosive materials, and explosives. Brief discussions of these categories follow. It should be noted that these categories are nonexclusive. For example, hydrogen peroxide is both a water reactive material and an oxidizing agent.

Water reactive materials may be hygroscopic, which means that their physical characteristics are significantly altered by the effects of water vapor. For example, a container of concentrated sulfuric acid may overflow when exposed to the water vapor present in the air in a classroom or laboratory.

Other water reactive materials may undergo hydrolysis, which is the decomposition or alteration of a chemical substance by water. Products of hydrolysis are often flammable or explosive. (In addition, they may also be toxic or corrosive.)

Water reactive materials deserve special attention apart from their inherent fire hazard properties as a flammable substrate. Initial flammability may be relatively insignificant compared with problems encountered in attempting to extinguish fires involving these materials. Water can react chemically with some water reactive substances to result in greater fire intensity than the original combustion. Because of this property of certain water reactive materials, precaution must be taken in selecting materials to fight fires. Table 2.1 presents a list of some common water reactive materials.

A substance that gains electrons in a chemical reaction is an oxidizing agent. Table 2.2 lists some common oxidizing agents in approximate order of oxidizing power. These substances tend to have oxygen in their structures, and often have names beginning with "per" prefixes (peroxide); and ending in "ate" (chlorate). Hazardous features of oxidizing agents include the following: They are chemically reactive, their reactions usually generate heat, and they may propagate fires by the release of oxygen. Fortunately, most fires involving these materials can be extinguished with water.

Organic compounds are associated with a number of potential hazards. Most are extremely flammable, and most are insoluble in water. Therefore, water may be ineffective as a fire extinguisher. Many organic compounds burn with very sooty flames, thus impeding the efforts of firefighters. In addition to fire danger-related hazards, numerous organic compounds have associated health hazards.

Polymers, which are generally organic compounds, commonly comprise resins, plastics, and fibers, and burn when exposed to high temperatures. Burning polymers often melt as they burn, and the resulting dripping material can spread fires. The surface of burning polymers usually chars, generating large amounts of smoke.

A corrosive material is one which undergoes a chemical process that converts minerals and metals into unwanted products. Acids and bases are common examples of corrosive materials. Acids react with metals, metallic oxides, and carbonates. Bases react with acids, metals, and salts. A

Table 2.1
Some Common Water Reactive Materials

Alkali Metals:	-Lithium
	-Sodium
	-Potassium
	-Cesium*
	-Rubidium*
	-Francium*
Other Metals:	-Magnesium
	-Zirconium
	-Titanium
	-Aluminum
	-Zinc
Hydrides:	-Diborane
	-Lithium Hydride
	-Sodium Hydride
	-Sodium Borohydride
Organometallic Compounds:	-Trimethylaluminum
	-Dimethylcadmium
	-Tetramethyltin
	-Tri (isobutyl)aluminum
Peroxides	
Nitrides	
Carbides (salt like)	
Phosphides	

* Little or no commercial use

substance may be corrosive as a liquid, a solid, or a gas. Corrosive materials can cause the ignition of organic materials on contact. Additionally, they may emit toxic gases.

An explosive is a chemical compound or mixture of compounds that suddenly undergoes a very rapid chemical transformation, with the simultaneous production of large quantities of heat and gases. The Department of Transportation (DOT) classifies explosives as Class A, B, or C. Class A explosives are materials, such as nitroglycerin, that are capable of detonation by means of a spark, flame, or shock. Classes B and C explosives are materials that ordinarily explode only under extreme conditions of temperature.

A special type of fire danger is posed by the phenomenon called BLEVE (pronounced bleh' vee). BLEVE is an acronym that stands for Boiling Liquid Expanding Vapor Explosion, which occurs when tanks containing flammable gases under pressure are exposed to fire. The combination of the

Table 2.2
Some Common Oxidizing Agents

Listed in decreasing order of oxidizing power*:

- -Fluorine
- -Ozone
- -Hydrogen Peroxide
- -Hypochlorous Acid
- -Metallic Chlorates
- -Lead Dioxide
- -Metallic Permanganates
- -Metallic Dichromates
- -Nitric Acid (concd)
- -Chlorine
- -Sulfuric Acid (concd)
- -Oxygen
- -Metallic Iodates
- -Bromine
- -Ferric Salts
- -Iodine
- -Sulfur
- -Stannic Salts

* Varies with condition, reaction, and form of chemical involved.

weakened structure of the tank and the buildup of internal pressure from the gas results in an instantaneous release and ignition of the vapor. The danger resulting from a BLEVE is immense. People have been killed and dismembered by fragments of exploded tanks, which can travel several hundred feet at great velocities. BLEVEs are often unexpected, as they are likely to occur 10 to 30 minutes after the tanks have been exposed to flames.

The following is a partial list of common compressed gases which may result in a BLEVE when exposed to external heat:

Propane

Butane

Butene

Ethylene

Hydrogen Sulfide

Isobutane

Isobutylene

Propylene

Vinyl Chloride

Methane

The first recommended response to a potential BLEVE is to evacuate the area, and the second is to cool the containers if it is possible to do so with minimum risk.

Cryogens

Cryogens are gases that must be cooled to less than −1500 °F to achieve liquefaction, and their hazardous properties involve the problems of handling extremely cold materials. Transportation and storage of such materials requires special attention. Cryogens have a high expansion rate upon vaporization, which may cause the rupture of their container. They have an ability to liquefy other gases. This property can result in an explosion if air is caused to solidify and block the venting tube of a storage container. Additionally, cryogens have a potential to damage living tissue, as contact with these materials in their liquid state will cause frostbite and deep skin-cell destruction.

Common cryogenic liquids include air, methane, nitrogen, argon, helium, carbon dioxide, hydrogen, and oxygen. Apart from their low temperature, several of these substances also present an asphyxiation hazard. They are generally packaged in high-pressure cylinders and escape in gaseous form if a leak occurs. If released into a confined space, they can reduce the oxygen content substantially. Exposure to such conditions can result in loss of consciousness and death by asphyxiation.

Toxicity

Pharmacology is the study of the nature, properties, and effects that chemicals have on living systems. Toxicology is a branch of pharmacology, and toxicity relates to a harmful effect on the body by physical contact, ingestion, or inhalation of a substance.

Paracelsus, the founder of modern pharmacology, wrote: "Poison is in everything, and no thing is without poison. The dosage makes it either a poison or a remedy." This is an apt description of how the toxicity of a substance is directly related to its concentration. Toxicity actually depends upon four factors:

1. The quantity of the material.
2. The rate and extent of absorption into the bloodstream.
3. The rate and extent of metabolism into by-products.
3. The rate and extent of excretion of the material or its by-products.

When mixtures of chemicals are involved, toxic effects may be increased through synergism. This phenomenon occurs when chemicals work together in such a way that they are far more toxic in combination than could be accounted for by simple additive effects.

Regardless of the relative toxicity of a hazardous material, no harm occurs without exposure to the substance. Exposure occurs when a substance is absorbed through the skin, ingested, or inhaled. The respiratory system is the most potentially hazardous route by which a substance can enter the body, since gases are absorbed into the bloodstream almost immediately upon inhalation. Gases considered to be hazardous include the following categories:

Anesthetic: material capable of causing a marked loss of muscular powers.

Asphyxiant: material capable of arresting respiration.

Irritant: material which produces local irritation or inflammation of eyes, skin, or respiratory membranes.

Smoke is a gas which commonly acts both as an asphyxiant and an irritant. Smoke poses a hazard in the following ways: It obscures vision, interferes with normal respiration, aids in the absorption of toxic gases, and can be carcinogenic. It is smoke, not burning, that is the cause of 80 percent of the fire-related deaths in the United States. Nearly two-thirds of those deaths occur away from the area that is burning.

Although toxicity is not a physical constant and, therefore, is difficult to measure accurately, many efforts have been made to quantify toxicity. Units by which toxicity is measured include Lethal Dose, 50 percent kill (LD50), and Lethal Concentration, 50 percent kill (LC50). Both are measurements of the amount of a substance which results in death for 50 percent of test specimens after a specific duration of exposure. The lethal dose is usually expressed in units of milligrams per kilogram (mg/kg), and lethal concentration is usually expressed in parts per million (ppm) by volume. The Threshold Limit Value (TLV) is the upper limit of a toxicant concentration to which an average, healthy person may be exposed all day, every day, without suffering adverse effects. TLV is usually expressed as ppm for gases in air and as milligrams (mg) or micrograms per cubic meter ($\mu g/m^3$) for fumes and mists in air.

One of the earliest documentations of health effects of hazardous materials was by a contemporary of Erasmus, who published a study of the diseases of metal miners in Central Europe in 1556. During the industrial revolution and continuing to the present, toxic effects in humans have been considered to be substantiated when a significant number of individuals develop symptoms or a disease related to a particular environment.

Exposure to hazardous materials is usually categorized as either acute or chronic. Acute refers to short-term, often one-time exposure, and chronic refers to long-term, repeated exposure. Chronic effects are more difficult to quantify than acute effects for these reasons:

1. *Chemical mixtures are involved.* Of utmost importance in evaluating toxicity is determining the effect of carcinogens, which are cancer-causing agents. Carcinogenic effects are difficult to document not only because cancer may not appear until many years after initial exposure, but because people are exposed to so many chemicals that it is difficult to separate and determine the effect of one specific substance.
2. *Duration and concentration of exposure must be determined.* This is usually no easy task. For example, researchers in Massachusetts are sorting through children's soiled diapers to determine whether children do, in fact, eat more dirt than adults. The results of the research (which is being conducted with the assistance of a forklift) are intended to be used to help set safety standards for levels of pollutants in soil.
3. *Small populations at risk make statistical analysis difficult.* Complicating this fact is the probability that by the time most "after-the-fact" exposure studies are undertaken, a significant portion of the subjects have moved and are no longer accessible.
4. *Psychophysiological complaints may mask quantifiable effects.* Psychophysiological complaints include those symptoms which are often impossible to attribute to a single physical cause: depression, irritability, dizziness, nausea, weakness, fatigue, insomnia, and numbness. Such stress-related illnesses may be difficult to separate from exposure results, especially where stress effects remain after actual exposure. This is discussed further in the section on stress.

In spite of these complications, deleterious health effects resulting from hazardous materials have been documented, and, in many cases, the nature of the physiological problems is consistent with the exposure effects from known contaminants. A complete understanding of the toxicity of hazardous materials undoubtedly will involve an analysis of a range of physiological systems, including genetic predisposition.

Radioactive Materials

Exposure to radioactive materials induces alterations in body chemistry, which, in turn, can lead to the ultimate destruction of an entire organ or death.

Radiation exposure causes physical changes that are generally associated with aging and increase the mortality rate of an exposed population. Various cancers are results of radiation exposure, as are certain genetic mutations. These effects of radiation can be classified as somatic and genetic effects.

Somatic effects are either early or delayed, with early effects observed when fairly large doses of radiation are received over the whole body in a matter of hours or less. Obvious immediate symptoms include nausea and vomiting. Depending on the severity of the exposure and the tolerance of the individual, this may be followed by severe blood changes, hemorrhage, infection, and death. Delayed somatic effects include leukemia, bone cancer, lung cancer, breast cancer, and various other types of cancer.

Genetic effects involve changes in chromosomes, which may be gene mutations or chromosome abnormalities. The harmful effects of gene mutations may be relatively minor, resulting in a slight increase in susceptibility to disease or a decrease of a few months in life expectancy. Other mutations may have much more significant effects, including death in the embryonic stage or infertility in offspring.

Determination of the effects of radiation depends on factors similar to those involved in toxic effects. The following must be taken under consideration: total dosage of ionization produced in the organism, density of the ionization, dosage rate, localization of the effect, and the rates of administration and elimination of radioactive material.

The unit of radiation dosage commonly used in measuring biological effects is the rem (roentgen equivalent man). The roentgen is a measure of the total amount of ionization that the quantity of radiation could produce in air. A biological dose, given in rems, is a measure of the biological damage to living tissue from the radiation exposure. A dosage in rems equals one in rads, multiplied by the relative biological effectiveness (RBE) of the radiation—a factor that reflects the effects of different densities of ionization along the path of radiation.

Table 2.3 lists recommended maximum radiation exposure limits. The total accumulated dosage in rems should not exceed the following:

$$\text{rems} = 5 \times (\text{age in years} - 18)$$

No accumulation of radiation is recommended for those under eighteen years of age. For occupationally exposed individuals, an additional guideline is a maximum of 3 rems over any period of thirteen consecutive weeks.

Stress Effects

As discussed previously, it is not always possible to separate symptoms of stress from the actual symptoms of toxicity. Various symptoms of central nervous system toxicity such as dizziness, fainting, seizures, blurred vision, depression, hyperactivity in children, suicides, suicide attempts, and nervous breakdowns are subjective to evaluate and may be aggravated by the stress and anxiety that normally occurs after an individual becomes aware of exposure to hazardous materials. However, some researchers in the field of effects of hazardous materials on human health feel deleterious phenomena can exist completely independent of conventional physical exposure.

Potential stressful situations resulting from nonphysical exposure to hazardous materials include lack of understanding the specific hazard

Table 2.3
Radiation Exposure Limitations*

Dose Limits for Occupational Exposure Combined Whole Body Occupational Exposure	
Prospective annual limit	5 rems in any one year
Retrospective annual limit	10-15 rems in any one year
Long term accumulation to age N years	(N-18) x 5
Skin	15 rems in any one year
Hands	75 rems in any one year (25/qrt)
Forearms	30 rems in any one year (10/qrt)
Other organs, tissues, and organ systems	15 rems in any one year (5/qtr)
Fertile women (with respect to fetus)	0.5 rem in gestation period
Dose Limit for Public or Occasionally Exposed Individuals	
Individual or occasional	0.5 rem in any one year
Students	0.1 rem in any one year

Source: Safety Engineering Standards, Industrial Indemnity Co., 1980

* Recommended by the National Council on Radiation Protection

presented by a specific chemical, general fear that the materials represent a long-term hazard which may surface long after actual exposure, and suspicion that information is being withheld by those in power.

Several practices can combat potential stress effects, and these become increasingly important the more routinely a worker is exposed to hazardous materials. Recommended measures include training programs, tracking physical baselines for future medical reference, establishing physical and psychological criteria for working with hazardous materials, evaluating personnel by established criteria, and developing personal programs for overall stress reduction.

ANALYTICAL CONSIDERATIONS

Any discussion of the properties of hazardous materials presupposes that some positive identification of the substance has been made. Since all identifications are ultimately based on some type of analytical procedure, it is appropriate to briefly consider the validity of such procedures. The most commonly used mechanism or protocol intended to insure the validity of analyses is a combination of quality assurance (QA) and quality control (QC).

QA is the testing and inspection of all or a portion of the final product to ensure that the desired quality level of product reaches the consumer. QC is the inspection, analysis, and action applied to a portion of the product or operation to estimate overall quality and determine what, if any, changes must be made to achieve or maintain the required level of quality.

For the purposes of this discussion, the terms "product" and "operation" may be assumed to refer to the analytical procedures used to identify and quantify hazardous materials. However, it is important to remember that QA/QC procedures may be applied to any product or operation, ranging from the preparation of permit applications to conducting a test burn for a hazardous waste incinerator.

QA programs are intended to assure the reliability of data and are implemented through written plans which usually include a description of QC measures. QC is thus a subset of QA. QA addresses the entire process of maintaining overall quality for a specific data-generating function and can include such processes as data review prior to analyses and post-analyses peer review. QC is usually the term applied to individual, specific acts within a given test, such as repetitive analyses. An established, routine QA program applied to each operation in the data-generating process can prevent the necessity of originating individual QC efforts for each process step.

The overall purpose of both QA and QC programs is to eliminate or reduce the occurrence of errors, which are the numerical differences between a measured value and the true value. Errors that can be ascribed to definite causes are termed determinate (or systematic) errors. Determinate errors are further classified as methodic, operative, or instrumental, depending on their identified source of origin. These refer to the method of analysis, capability of the operator, and mechanical failures of measuring equipment, respectively. Frequently, the source of error may exist in more than one category.

Indeterminate errors are a more philosophical concept than determinate errors. This term describes the phenomena where unpredictable factors introduce what appear to be random fluctuations in the measured quantity. An awareness of the existence of indeterminate error is valuable in data evaluation.

Figure 2.1 depicts the flowchart of a sequence of events during a controlled series of analytical measurements. Viewed in this manner, it is easy to identify numerous points where determinate error might be introduced into the analytical procedure, including events prior to the actual analyses.

Assuming some type of sample is to be analyzed, it is appropriate to review the definition of sample: groups or portions of material, taken from

Figure 2.1
Flowchart of the Sequence of Events During a Controlled Series of Laboratory Measurements

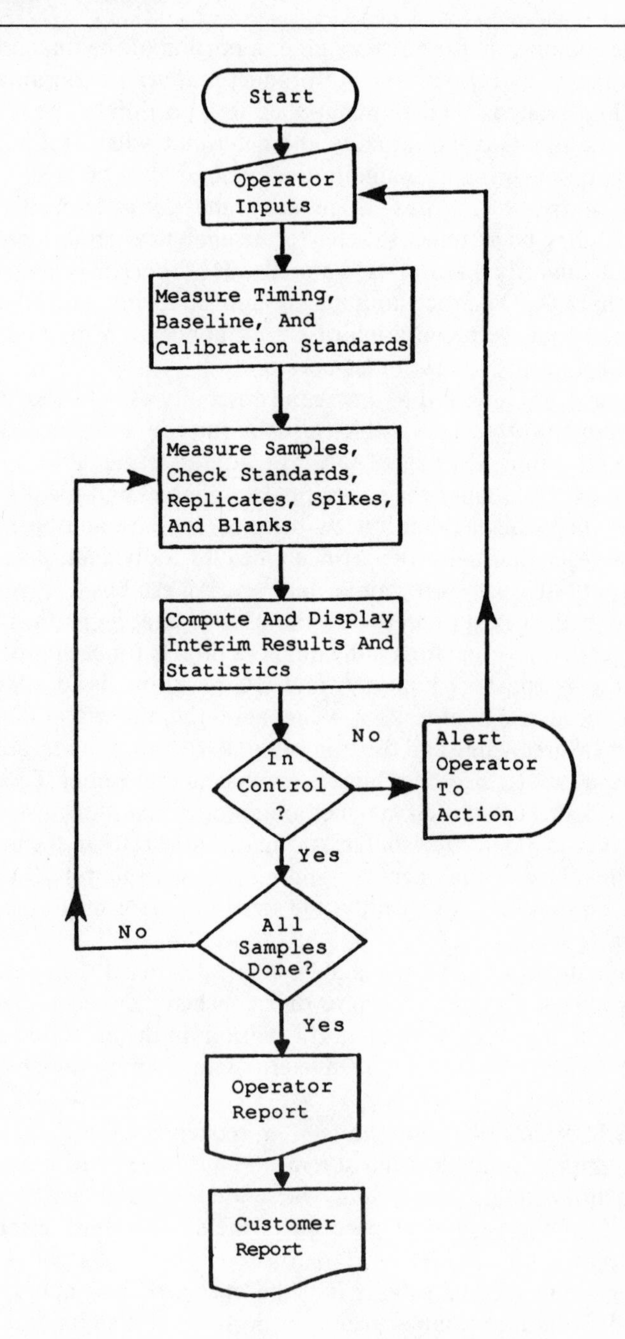

a larger collection of units or quantity of material, that provide information to be used for judging the quality of the total collection or entire material as a basis for action. Error can be introduced into an analysis if a sample is nonrepresentative or has been contaminated in some way.

The potential for error introduction continues through the analytical process to the final reporting of the results. To obtain meaningful data, the analyst must perform the proper analysis in the prescribed fashion, complete required calculations, and convert the results to final form for permanent recording of the analytical data in meaningful, exact terms. In addition to analytical procedures, good quality assurance for environmental measurements includes assessments of the following: detection limits of analytical methods, data sampling procedures, sample handling and storage, calibration, documentation, and evaluation of appropriateness of measurement method. Therefore, effective QA/QC programs address not only the analytical procedures, but envelop the entire analytical activity from inception of sampling plans through reporting procedures.

It is highly desirable that analytical results be both accurate and precise. Accuracy is the extent to which the results of a calculation or the readings of any instrument approach the true values of the calculated or measured quantities and are free from error. Precision is the quality of being exactly or sharply defined or stated (i.e., precise results are reproducible). Figure 2.2 graphically depicts the differences between accuracy and precision.

Spikes and dupes are often used to evaluate accuracy and precision of analytical procedures. A spike is a sample to which a known amount of a specific substance has been added. A dupe, also known as a replicate, is one of two or more samples which contain the same materials in identical amounts. With spikes, the accuracy of the analytical results can be compared with the "right" answer. The precision of analysis can be evaluated by determining how closely the results for duplicated samples (dupes) match.

Analyses to evaluate the quality of laboratory procedures are often conducted "blind" by a facility with third-party status. This means that the individuals who conduct the analyses have no information about the content of the samples and have no preconceived ideas or vested interest regarding the results.

SUMMARY

Today's technologies use materials that were unknown as recently as 50 years ago. Many of these materials have hazardous properties, as discussed in detail in this chapter. They may burst into flame, explode, or cause illness or death. An intelligent awareness of the unique requirements for handling hazardous materials goes hand in hand with proper identification of specific materials and an understanding of their physical and chemical properties.

Proper identification must be coupled with an awareness of the limita-

Figure 2.2
Accuracy and Precision

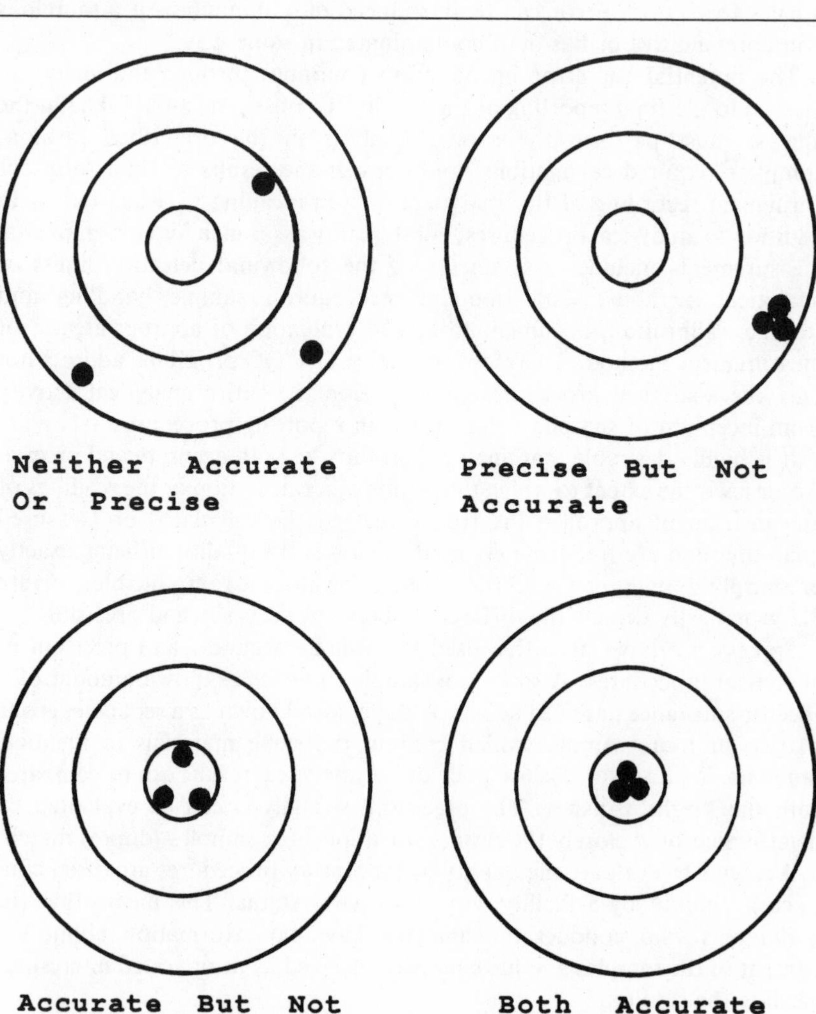

tions of any analytical procedure. As stated in *Quantitative Analysis* (Prentice-Hall, 1967), "The true value of any quantity is really a philosophical abstraction, something that man is not destined to know, although scientists generally feel there is such a thing and believe that they may approach it more and more closely as their measurements become increasingly refined."

3

THE RULES OF THE GAME: REGULATIONS

> One of the things concerning us . . . is how one turns science into regulation, or how one turns what is known or not known in the scientific community into regulations that protect the health and safety of individuals and the environment.
>
> —Alan McGowan, President
> Scientists' Institute for Public Information

Environmental law consists of all the legal guidelines that are intended to protect our environment. Such directives may originate from the United States Constitution; state constitutions; federal and state statutes and local ordinances; regulations promulgated by federal, state, and local regulatory agencies; court decisions interpreting these laws and regulations; and/or common law.

Simply defined, a law is a statement that tells a segment of society what it must do. Various agencies of government may then prepare regulations, which define how an activity must be conducted in order to comply with the law. Therefore, regulations are prepared in response to, and under the direction of, laws. Regulations then are part of the law and have the full force of the law. In practice, however, the terms "law," "statute," and "regulation" are often used interchangeably.

However, in order to obtain copies of laws or regulations, it is important to differentiate between the statutes (laws) and the regulations. Laws can be accessed through their Public Law number from the U.S. Government Printing Office and are compiled in the United States Code (USC). Regulations are generally more volatile than laws, of more applicability in determining compliance, and are generally the documents most frequently consulted in hazardous materials management, as opposed to statutes or laws.

Regulations are printed in the *Federal Register* (FR) and are compiled annually in the *Code of Federal Regulations* (CFR).

CFR Title 29 contains regulations mandated by the Occupational Health and Safety Administration (OSHA), Title 40 contains EPA regulations, and the Department of Transportation (DOT) regulations are found in Title 49. The U.S. Nuclear Regulatory Commission (NRC) regulations appear in Title 10.

Guidelines on how to use the CFR are provided within each volume. On the cover are listed the number, parts included, and revision date. An "Explanation" section at the front of each CFR volume lists information such as issue dates, legal status, how to use the CFR, and so on. More detailed information on using the CFR is included at the end of each volume. The "Finding Aids" section is composed of the following subsections:

1. Materials Approved for Incorporation by Reference;
2. Table of CFR Titles and Chapters;
3. Alphabetical List of Agencies Appearing in the CFR;
4. Appendix to List of CFR Sections Affected; and
5. List of CFR Sections Affected.

Regulatory programs under the general environmental laws are developed through the process of administrative rule making. In most cases, the Administrative Procedure Act (APA) is the controlling statute for rule making. Except for statutes requiring formal hearings, APA provides for informal rule making: The involved agency is required to publish a notice of the proposed rule and allow interested parties to submit written comments. Hearings are not required. After considering all submitted comments, the involved agency prepares and publishes a justification supporting the final rule and issues the final rule. Because the majority of federal environmental laws are formulated by EPA, a discussion of that agency's rule making procedure follows and is graphically depicted in Figure 3.1.

EPA's informal rule making usually begins when proposed rules are published in the FR and comments are requested. In certain instances an Advance Notice of Proposed Rulemaking (ANPR) or position paper may precede publication of the proposed rule, but this is not a procedural requirement. A period of 60 days is normally provided for receipt of comments, but the comment period can be shorter or longer. In addition to written comments, there may be testimony presented at public meetings, hearings, and technical conferences. Input is generally available for review under the Freedom of Information Act. At the end of the comment period, EPA analyzes the comments which have been received and determines what, if any, changes are appropriate to make in the proposed rule.

Figure 3.1
The Process of Rule Making

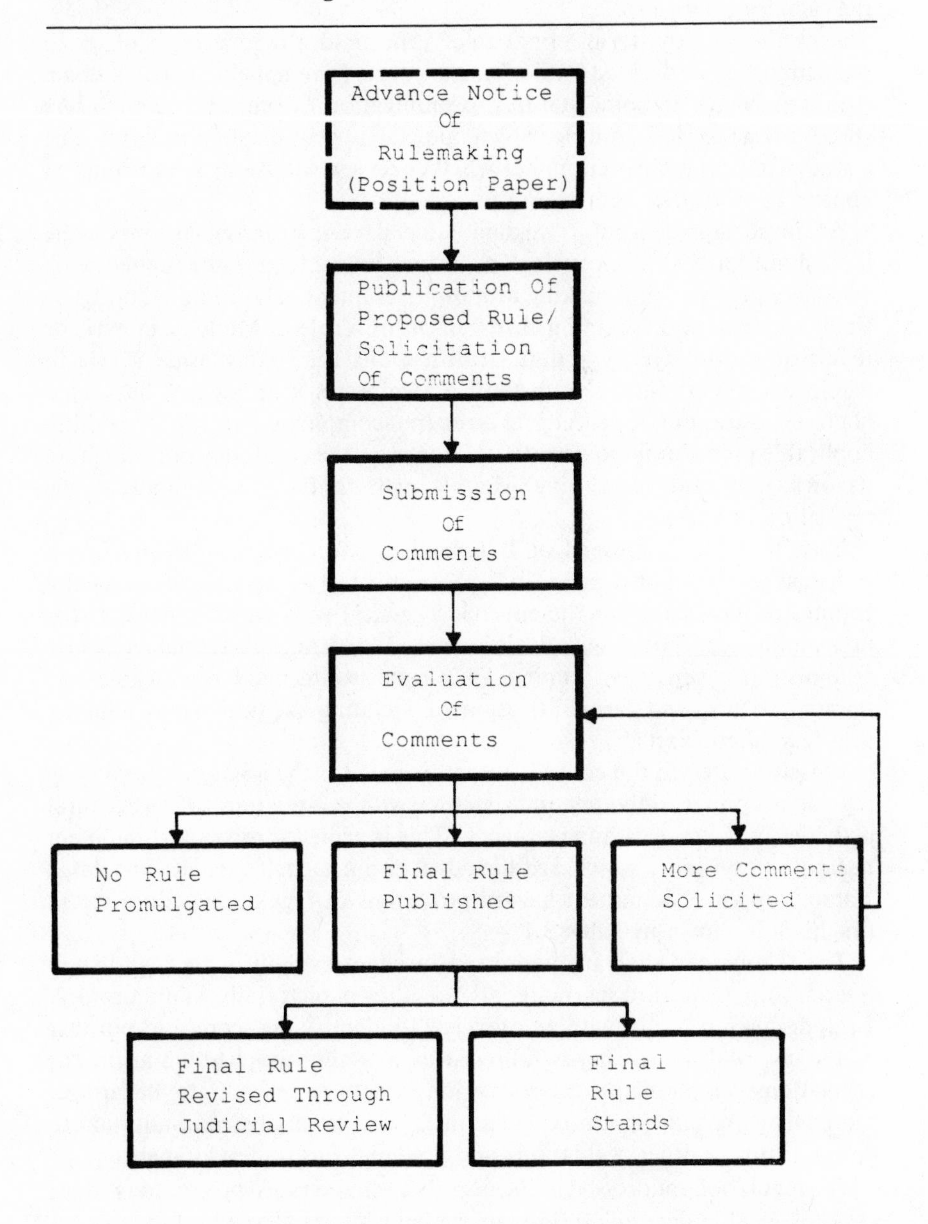

The final rule is then published in the FR, unless it has been decided that no rule will be published or an interim final rule is published. An interim final rule serves as a mechanism to allow further comment and modification. These rules are enforceable and generally subject to judicial review—the same as final rules.

Regardless of the term "final rule," the final effective form of many regulations is not decided until after the courts have upheld or struck down their provisions. In some statutes, promulgation of the final rule triggers specific time periods during which judicial review must be sought. The statute itself generally determines whether review will occur in the court of appeals or in district court.

An important concept regarding judicial review of regulations is the limited authority of the courts. The agency promulgating the regulation is considered the decision-making arm of government. The court is limited to finding against and enjoining enforcement of a rule in whole or in part, or to affirming the agency action. Justifications for setting aside a rule in whole or in part include substantive legal error (the agency has acted contrary to statute) or procedural error (noncompliance with APA or other applicable procedural requirements). However, the court may not substitute its own judgment concerning what "ought to be" the content of the regulation in question.

Even if never challenged or if upheld in court, regulations are of no practical use without a means of enforcement. The major environmental statutes provide EPA (or the enforcing agency) with various remedies for noncompliance. These include injunctions, orders, substantial civil and administrative penalties, criminal fines, imprisonment of responsible corporate officers, and disqualification of violating facilities from entering into federal contracts.

Closely related to the concept of enforcement is the ability to invoke the regulatory powers inherent in a statute with a minimum of procedural activities in the case of an emergency. This is achieved through "imminent hazard" provisions, which are included in most environmentally related statutes. A list of imminent hazard provisions of regulations addressed in this book is shown in Table 3.1.

The regulations that are addressed in chapters 4 through 6 and their relationship to hazardous materials are shown pictorially in Figure 3.2. Each discussion begins with an overview to identify the scope and purpose of the law. A brief history is followed by a section which summarizes the regulations, emphasizing those portions that are relevant to hazardous materials management. Enforcement measures are discussed in addition to relationships to other regulations and potential future developments.

The regulations addressed in the next three chapters are presented in loose chronological order and reflect an evolving national philosophy toward

Table 3.1
Imminent Hazard Provisions of Federal Statutes

Statute	Location in Statute	Location in U.S. Code
Comprehensive Environmental Response, Compensation, and Liability Act ("Superfund")	SS104,106	42 U.S.C. S9604,9606
Resource Conservation and Recovery Act	S7003	42 U.S.C. S6973
Clean Water Act	S311(e)	33 U.S.C. S1321(e)
Clean Air Act	S303(a)	42 U.S.C. S7603(a)
Safe Drinking Water Act	S3001	42 U.S.C. S300(i)
Toxic Substances Control Act	S7	15 U.S.C. S2606
Hazardous Materials Transportation Act	S111	49 U.S.C. S1810(b)
Occupational Safety and Health Act	S13(a)	29 U.S.C. S662(a)
Federal Insecticide, Fungicide, and Rodenticide Act	S6(c)	7 U.S.C. S136(e)

environmental protection. For the purpose of a conceptual visualization of regulatory development, the regulations may be grouped as shown in Table 3.2. Group 1 is composed of pesticide regulations, occupational safety regulations, and national environmental policy. This group resulted from, or constituted, strong historical precedent. The second group of regulations, promulgated in the 1970s, consists of laws specifically intended to protect air and water quality. The last group consists of the most recently enacted regulations—those that specifically and in their entirety address hazardous and toxic materials management.

Asbestos and radioactive materials control are discussed in chapters 7 and 8. The regulation of these materials is governed by a compendium of laws administered by various agencies.

Figure 3.2
Hazardous Materials in the Regulatory Framework

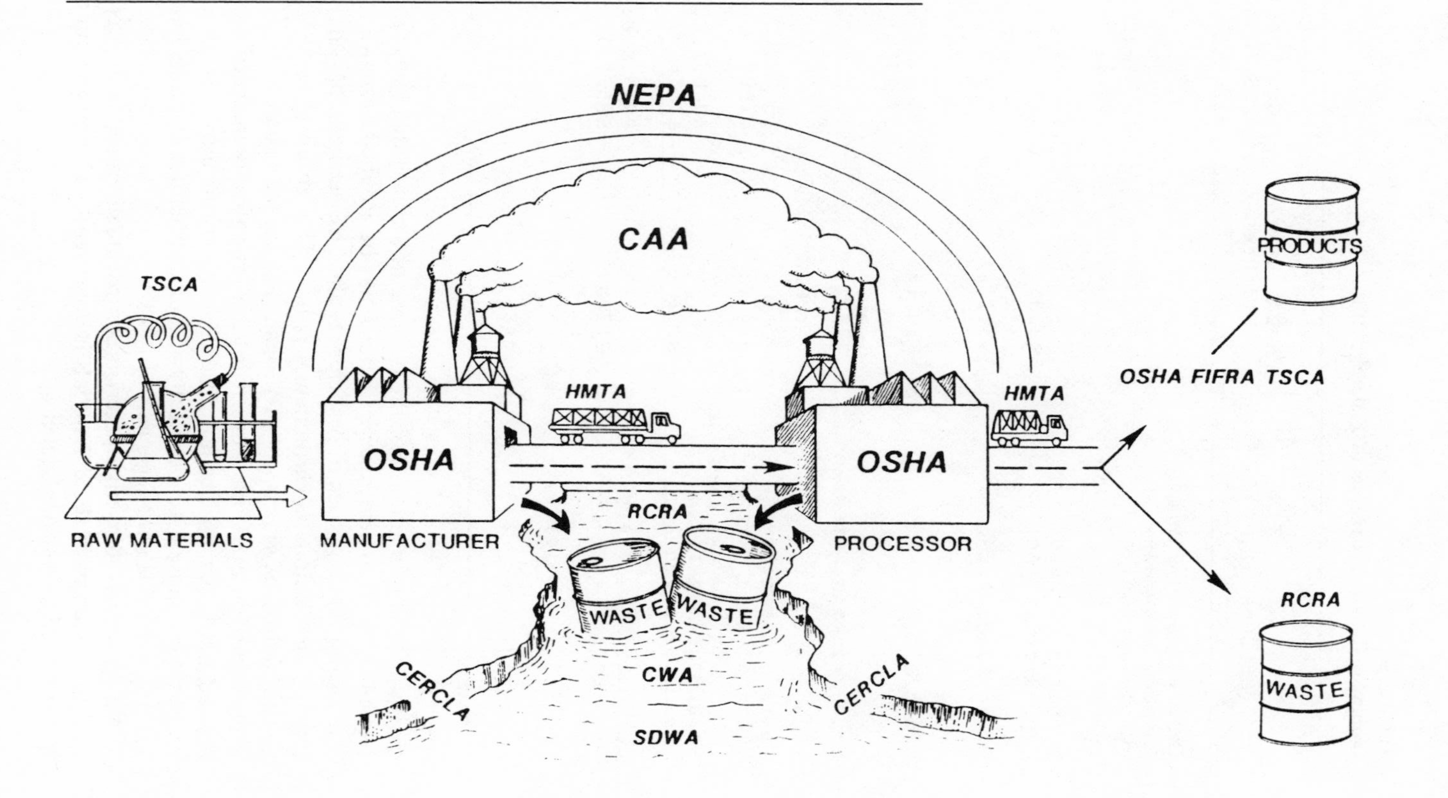

Table 3.2
Federal Laws Addressing Hazardous Materials Management

Group Identifier	Regulation	Year 1st Enacted	Agency in Charge	Sources Covered
Legislation addressing long-standing environmental concerns	Federal Insecticide, Fungicide & Rodenticide Act (FIFRA)	1947	EPA	Pesticides
	Occupational Health & Safety Act (OSH Act)	1970	DOL (OSHA)	Chemicals in the workplace
	National Environmental Policy Act (NEPA)	1970	EPA	Federal actions with environmental impacts
Air and water quality regulations	Clean Air Act (CAA)	1970	EPA	Air pollutants
	Clean Water Act (CWA)	1972	EPA	Water pollutants
	Safe Drinking Water Act (SDWA)	1974	EPA	Drinking water contaminants
Toxic/hazardous substances regulations	Hazardous Materials Transportation Act (HMTA)	1975	DOT	Transportation of hazardous materials
	Toxic Substances & Control Act (TSCA)	1976	EPA	Chemical substances manufacturers, distribution
	Resource Conservation & Recovery Act (RCRA)	1976	EPA	Solid and hazardous waste
	Comprehensive Environmental Response, Compensation & Liability Act (CERCLA)	1980	EPA	Hazardous waste spills and abandoned or inactive hazardous waste sites

The laws discussed in the following chapters are federal laws. It is important to remember that states and other local governments can add another layer of regulation. Many environmental acts encourage state participation provided that the state plans meet or exceed federal guidelines. On the other end of the spectrum, local governments may be barred by certain acts from enacting legislation which is more stringent than the federal regulations.

Also, the laws discussed in the following chapters are by no means the only federal laws that address hazardous material. Table 3.3 is a list of additional laws that may also regulate certain hazardous materials.

Environmental regulation is far from a static subject. Numerous changes occur because of legal interpretations by the courts, revisions by the issuing agency, and the outcome of ongoing environmental legislation. Statutes may be revised with such frequency that confusion occurs as to what is officially contained in the provisions at any one time. Therefore, when researching a specific statute in detail, the reader should take care to verify the status and content of related regulations at any point in time.

Even then, such knowledge may be like trying to pinpoint a moving target. For a comprehensive approach to becoming knowledgeable about a regulation, it is also important to know what current judicial trends are, the

Table 3.3
Other Laws Affecting Hazardous Materials

- Consumer Product Safety Act
- Dangerous Cargo Act
- Deepwater Ports Act*
- Federal Disaster Relief Act
- Federal Environmental Pesticide Control Act
- Federal Food, Drug, and Cosmetic Act
- Federal Hazardous Substances Act
- Federal Mine Safety and Health Act
- Federal Railroad Safety Act
- Flammable Fabrics Act
- Intervention on the High Seas Act*
- Lead-Based Paint Poisoning Prevention Act
- Marine Protection, Research, and Sanctuaries Act*
- Outer Continental Shelf Lands Act*
- Pipeline Safety Act*
- Poison Prevention Packaging Act
- Ports and Waterways Safety Act*
- Trans-Alaska Pipeline Act*
- Uranium Mill Tailings Radiation Control Act

* Primarily relate to oil and fuels as a hazardous substance.

status of budgetary controls and how they can reduce or broaden an act's impact, how Congress exerts its controls via appropriation review, the status of proposed amendments, and the sentiments of the public.

The field of environmental regulations undeniably involves a rare combination of science, technology, and law. The qualities of a successful soothsayer, able to see into the future, would undoubtedly be a helpful addition to more conventional backgrounds in these fields.

4

HISTORICAL PRECEDENT REGULATIONS

It is the continuing policy of the Federal Government . . . to create and maintain conditions under which man and nature can exist in productive harmony . . .

—Section 101(a)
National Environmental Policy Act, 1970

Pesticide regulations and occupational safety laws resulted from long-standing safety concerns, with popular support established in the 1800s and earlier. Both later became subjects addressed under the broader topic of environmental issues.

While such specific concerns were resulting in federal legislation, Congress was becoming receptive to a widespread desire on the part of the nation to address environmental quality in a unified, comprehensive manner. The National Environmental Policy Act (NEPA) was the legislative response to an awakening awareness of the temporal nature of environmental quality.

Whereas pesticide control and occupational safety had as precedents an established base of public support, NEPA itself established an important precedent. There is no express mandate in the Constitution directing the government to protect the environment, just as there are no directives addressing health and safety. Constitutional provisions addressing these issues are interpreted by Congress to be implied rather than express, drawing heavily upon the preamble in the Constitution which offers the promise to "promote the general welfare."

Implied justification for the regulations discussed in this book rests in part on this provision and the congressional right to regulate affairs affecting commerce. For environmental regulations which were promulgated after NEPA, the act itself provided supporting justification.

THE FEDERAL INSECTICIDE, FUNGICIDE, AND RODENTICIDE ACT (FIFRA)

FIFRA is intended to act as a guideline in providing protection for persons, animals, organisms, and environments which come into contact with pesticides. The act regulates the production, packaging, marketing, transportation, application, and effects of pesticides in relation to possible adverse conditions such activities may create. For the purposes of this regulation, a pesticide is broadly defined as almost anything intended to be used to kill, repel, or control any nonhuman form of life. Microorganisms or drugs that affect man or animals are excluded from regulation by FIFRA.

Under FIFRA, the Office of Pesticide Programs (OPP) of EPA is responsible for registering new pesticides, new uses of existing pesticides, and reregistering currently registered pesticides. Whether a pesticide should be registered or reregistered is based on a review of health, environmental, and economic information. From this information, the OPP weighs the risks of use against the benefits of use. If the benefits outweigh the risks, the OPP can register or reregister the pesticide.

FIFRA identifies enforcement activities such as pesticide regulation through registration, analysis, inspection, recordkeeping, and administrative review. In addition, FIFRA identifies corrective and/or punitive action which may be taken in the event of regulation violations.

The hotline number for information on pesticide-related questions is 1-800-858-7378.

Background

The need for pesticide regulation is obvious upon consideration of pesticide/herbicide quantities and characteristics. These substances, among the most toxic produced by the chemical industry, are manufactured in numerous and varied quantities. Currently, over 40,000 pesticide products are made from one or more basic chemical compounds and more than 1,200 active pesticide ingredients. Table 4.1 lists characteristics and trade names of major types of synthetic pesticides.

Today's FIFRA is the outgrowth of some of the first laws in the United States to address a type of environmental protection—a concern with the proper use of pesticides. The insecticide known as Paris Green was used as early as 1865 in the United States to control the Colorado potato beetle. In 1910, Congress first set standards for ingredients in selected pesticides, including Paris Green and lead arsenate.

The original FIFRA was adopted in 1947 as a guideline for pesticide registration and labeling, due to widespread increase in domestic use of synthetic organic pesticides following World War II. Theoretically, the act had established three review points for a pesticide; however, as administered by the

Table 4.1
Common Pesticides

Chemical Components	Characteristics	Names
Chlorinate hydrocarbons (organochlorines)	Toxic, persistent	DDT, Chlordane DDD, Dieldrin, Aldrin, Endrin, DDE, Toxaphene, Heptachlor, Lindane
Organic phosphates (organophosphates)	Toxic, degradable	Parathion, Malathion, Chlarethion, Thimet, Methyl-parathion, Phosdrin, Trichlorphone
Carbamates	Low toxicity to humans	Sevin, Snip, Baygon, Ficam

U.S. Department of Agriculture (USDA) there was actually no substantial safety review. Therefore, in the 1960s, the courts gave the USDA and subsequently EPA the discretion to ban pesticides based on a risk/benefit comparison. (With the establishment of EPA in 1970, administration of FIFRA was transferred from the USDA to EPA.) Amendments made in subsequent years, and the supplementary Federal Environmental Pesticide Control Act of 1972 (FEPCA), further clarified the objectives of the act in regulating the use, handling, and production of pesticides.

Content

The regulations comprising FIFRA embody the following four goals:

1. Evaluation of risks using a *registration* system;
2. Exposure control through *classification* and *certification*;
3. *Suspension*, *cancellation*, or *restriction* of pesticides that pose a risk to the environment; and
4. *Enforcement* through inspections, labeling notices, and regulation by state agencies.

FIFRA requires that all pesticides be registered according to regulation prior to shipment, delivery, or sale in the United States. The risks to the environment from pesticide use must be evaluated by EPA as part of the registration procedure. EPA has the authority to require manufacturers to submit relevant data deemed necessary to decide whether or not to register a pesticide. Evidence for registration must demonstrate that the product will not cause unreasonable harm to humans, that it will perform its intended function without adverse effects, and that it will not result in harmful residues on food or feed.

In addition to registration of pesticides, each pesticide-producing establishment must apply for registration and obtain an EPA-issued establishment registration number. This number does not replace the product registration number: Both must be shown on the pesticide label or container. Registered establishments must submit annual reports to EPA.

In addition to the standard registration processes, pesticides can be distributed and used under limited registrations or exemption mechanisms. These special case-by-case approvals include conditional registrations, state registrations for special local needs, emergency uses for specific problems, and experimental use permits.

In registering a pesticide, EPA must classify the substance as being for general or for restricted use. This requirement became effective in 1977. Most pesticides are classified for general use, which means that the product may be sold with no restrictions on who uses it. Pesticides classified for restricted use may be used only by, or under, the direct supervision of a certified applicator. These restricted substances are usually extremely toxic or for some other reason present a special hazard.

Certification programs for applicators are generally administered by state agencies. These programs are monitored by the federal government, which provides funding assistance in certain circumstances. There are two basic classes of pesticide applicators: private and commercial. A private applicator may use or supervise the use of restricted-use pesticides to produce any agricultural commodity on property owned by the applicator or the applicator's employer, or without compensation on another person's property. A commercial applicator is certified to use or supervise the use of restricted-use pesticides for any purpose on any property.

EPA may suspend, cancel, or restrict a current pesticide registration if the substance is deemed to pose an unreasonable risk to humans or to the environment. Instead of invoking such stringent measures, the agency has more commonly dealt with potentially hazardous pesticides through an extended special review process, termed Rebuttal Presumption Against Registration (RPAR). A RPAR can result in cancellation of all or some registered uses, reclassification, or other regulatory restrictions. However, a manufacturer may respond to a RPAR by trying to rebut EPA's final decision.

During the 1980s, the practice of substituting a "Special Pesticide Review" for suspect chemicals to replace RPARs has been widely adopted by EPA. This practice has proven generally popular with both industry and the regulatory agencies, as a RPAR is an expensive and time-consuming procedure. By 1984, the backlog of RPAR reviews in progress had been reduced to nearly zero.

Control of pesticide labels is a key enforcement tool under FIFRA, enabling EPA to control and restrict the use of pesticides. It is a violation to sell a pesticide with a label that does not meet EPA standards. Further, it is a violation to use a pesticide in a manner that is inconsistent with directions specified on the label.

Pesticides classified as highly or moderately toxic must bear the words "Danger," "Poison," or "Warning" and/or the skull and crossbones symbol on the label. Others must bear the word "Caution." EPA began a label improvement program (LIP) in 1980 intended to upgrade the extent of information presented on pesticide labels. Areas targeted for expansion and clarification include storage, disposal, and farmworker safety information.

Enforcement actions against violations range from fines to civil or criminal penalties. FIFRA allows the responsible agency to undertake the following actions in order to enforce the law:

1. Register facilities involved in the manufacture, formulation, or distribution of pesticides;
2. Inspect production facilities;
3. Examine and test pesticides offered for sale;
4. Impose fines or criminal penalties for violations; and
5. Stop sales of, seize, or recall products violating the regulations.

Enforcement

FIFRA is enforced by EPA through the Pesticides and Toxic Substances Enforcement Division. This division works in conjunction with the OPP in monitoring the conduct of pesticide enforcement and registration programs.

The division makes decisions concerning suspension or cancellation of registrations and determines when additional enforcement actions are appropriate. In addition, the division reviews scientific analyses of pesticide samples and makes recommendations to aid in achieving compliance; coordinates recall and seizure activities when necessary; and pursues legal action, both civil and criminal, when appropriate.

Individual states may have primary enforcement responsibility granted that the state has adequate pesticide use laws, does not permit any sale or use of pesticides prohibited by FIFRA, does not impose any labeling or packaging requirements different from those established by EPA, has adopted and is implementing effective enforcement procedures, and maintains records in accordance with the federal guidelines. All 50 states currently have functioning FIFRA programs.

EPA may not require a state to have pesticide use laws and regulations more stringent than FIFRA. To date, states may, however, ban or limit use of specific pesticides and may require manufacturers to submit additional data to obtain state registration.

Relationship to Other Regulations

Spills of pesticides are subject to federal regulations in the Clean Water Act (CWA) covering discharges of toxic and hazardous materials into water, and pesticide manufacturers must apply for discharge permits if they release effluents into any regulated body of water. Pesticides in the atmosphere may be regulated by the section of the Clean Air Act (CAA) which pertains to hazardous air pollutants. The Resource Conservation and Recovery Act (RCRA) addresses the disposal of pesticides and wastes from pesticide manufacturing processes. Where appropriate, testing required by the Superfund Amendments and Reauthorization Act (SARA) will be carried out under authority granted by FIFRA.

The U.S. Department of Labor (DOL) has responsibility for the protection of workers from pesticide hazards through administration of the Occupational Safety and Health Act (OSH Act).

Potential Future Developments

Because some states have promulgated guidelines for pesticides, which are more restrictive than federal regulations, legislation has been proposed to limit state authority to gather data about a pesticide for state registration. Also, measures to protect farmworkers exposed to pesticides and more stringent regulations on imported and exported chemicals have been proposed at various times.

In recent years, legislation introduced to amend FIFRA has failed to pass, as pesticide law does not appear to be a high congressional priority. Other major environmental bills have blocked the passage of proposed amendments to FIFRA. With congressional resolution of competing major environmental legislation, future legislative sessions may result in amendments to FIFRA.

At the agency level, more stringent rules concerning the use of pesticides in relation to farmworker protection and threatened or endangered species are likely to be implemented in 1988. EPA continues to study the use of three controversial pesticides—chlordane, heptachlor, and aldrin—and has requested manufacturers to tighten use restrictions. Citing a study that demonstrates significant cause for concern related to chlordane use, EPA has announced its intention to decide on a new regulatory scheme for this substance.

OCCUPATIONAL SAFETY AND HEALTH ACT (OSH ACT)

The purpose of the OSH Act is to assure safe and healthful employment conditions for all men and women working in the United States. The act addresses the proper management of hazardous materials in the working environment by setting standards that are intended to prevent injury and illness among workers. Employers are required to inform workers of the dangers posed by substances with which they work. An additional goal of the act is the definition of a standard policy for regulating substances which cause cancer.

Regulations defined by the OSH Act are commonly associated with accident prevention, which was a driving impetus in promulgating the original legislation. However, since its enactment, the focus of the OSH Act has increasingly concentrated on the proper management of hazardous materials. The rationale behind this trend is evident upon study of Table 4.2. The ten leading job-related diseases and injuries have been identified by the National Institute for Occupational Safety and Health (NIOSH) and are listed in rank order in the table. Only two items on this list, numbers two and four, are commonly due to traumatic accidents. With the exception of item 8, the remainder normally result from exposure to toxic or hazardous materials. In light of the estimate that one in four workers is exposed to one or more chemical hazards, the progression to a focus on hazardous materials is understandable.

Table 4.2
Leading Work-Related Diseases and Injuries

Ranked in order of incidence and severity:

1 - Occupational lung diseases
2 - Musculoskeletal injuries
3 - Occupational cancers
4 - Amputations, fractures, eye loss, lacerations, and traumatic deaths
5 - Cardiovascular diseases
6 - Disorders of reproduction
7 - Neurotoxic disorders
8 - Noise-induced hearing loss
9 - Dermatologic conditions
10 - Psychologic disorders

Background

A concern with the assurance of worker safety has long been a governmental issue, both at the state and federal levels. In 1877, Massachusetts passed the first statute intended to insure worker safety. In 1910, in response to countless mining disasters, Congress created the Bureau of Mines in an effort to address the safety concerns involved with one high-hazard occupation. The current OSH Act was enacted in December 1970, in the same month that EPA was created.

To a great extent, the groundwork for the act was laid as part of the Great Society programs of President Johnson, in response to numerous testimonies alleging that working conditions in U.S. industries were hazardous and costly in terms of time lost from work due to illnesses and injuries. A 1968 legislative effort to address worker safety was unsuccessful, but the need for this type of federal law had been established. The Senate and House passed separate bills on worker safety in 1970. These were resolved into a compromise measure which became the current OSH Act.

Content

The act establishes conditions that call for the following actions:

1. Encouraging employers and employees to establish or perfect programs intended to reduce occupational hazards;
2. Providing that employees and employers have separate but dependent rights and responsibilities in relation to achieving safe and healthful working conditions;
3. Authorizing the Secretary of Labor to establish mandatory guidelines for occupational safety and health applicable to business affecting interstate commerce;
4. Creating an Occupational Safety and Health Review Commission for carrying out various functions defined by the act;
5. Augmenting previous advances made toward assuring occupational safety and health;
6. Providing means for research in the field of occupational safety and health;
7. Exploring ways of identifying occupation-related latent diseases;
8. Establishing medical criteria intended to assure (within reasonable limits) that no employee suffers diminished health, functional capacity, or life expectancy as a result of working conditions;
9. Providing for training programs to expand the field of occupational safety and health;
10. Providing for the development and enactment of occupational safety and health standards;
11. Providing an effective enforcement program which prohibits providing advanced warning to those responsible for areas to be inspected or investigated;
12. Providing grants to the states to encourage governments to expand, administer, and enforce state occupational safety and health laws;
13. Providing for appropriate reporting and record-keeping procedures in regard to occupational safety and health; and
14. Encouraging joint efforts to reduce injuries and diseases that result from working conditions.

The Occupational Safety and Health Administration (OSHA) is charged with administering the provisions of the OSH Act, which is generally applicable to facilities with ten or more workers. This usually excludes farmworkers and the self-employed from regulation under the act. Federal employees and workplaces regulated by other federal agencies are also exempt, except for the Hazard Communication Standard. The major areas where OSHA regulations address hazardous materials are the following: health standards, cancer policy, the Hazard Communication Standard, and OSHA's Hazardous Waste Operations and Emergency Response rules.

Section 6 of the OSH Act grants authority to set safety and health standards. There are three types of standard setting procedures: consensus, permanent, and emergency temporary.

Consensus standards are those which were adopted during the period of 1971-1973 from other federal agencies, various industries, or private organizations. These resulted in a preliminary initial package of standards addressing several hundred toxic chemicals, with threshold exposure values as the only elements in the standard. Some of these substances have subsequently been addressed in permanent standards. Others have since been revoked by Congress.

Permanent, or final, standards are usually developed by NIOSH, which was created to conduct research on safety and health issues and is charged with the development of criteria documents which are to be used as a basis for preparing new standards or for recommending more complete standards. These documents are prepared from a critical evaluation of all published information and data. Related federal agencies that may contribute data include the National Cancer Institute; National Heart, Lung and Blood Institute; National Institute of Environmental Health Sciences; National Center for Health Statistics; Centers for Disease Control; and National Center for Toxicological Research.

Asbestos was the first chemical covered by a specific final standard issued in 1972. By 1985, OSHA had also promulgated final standards for the following substances: vinyl chloride; inorganic arsenic; lead; coke oven emissions; cotton dust; 1, 2-dibromo-3-chloropropane; acrylonitrile; coal tar pitch volatiles; benzene; and cotton gin dust. However, the standards for benzene and cotton-ginning dust were subsequently overturned by the courts.

Specific standards vary in form and requirements. In general, however, each addresses exposure limits, labeling, protective equipment, control procedures, monitoring employee exposure, measuring employee exposure, medical examinations, and access to records. Probably the most important parameter addressed in standards is exposure. Terms commonly used by NIOSH to describe exposure limits include the following:

—Action Level: the exposure concentration at which certain provisions of the NIOSH-recommended standard must be initiated, such as periodic measurements of worker exposure, training of workers, and medical surveillance.

—Ca: NIOSH recommends that the substance be treated as a potential human carcinogen.

—Ceiling: a description usually seen in connection with a published exposure limit; referring to a concentration that should not be exceeded, even for an instant.

—Immediately Dangerous to Life or Health (IDLH): level defined only for the purpose of respirator protection, representing a maximum concentration from which, in the event of respirator failure, one could escape within 30 minutes without experiencing any escape-impairing or irreversible health effects.

—Permissible Exposure Limit (PEL): an exposure limit that is published and enforced by OSHA as a legal standard.

—Recommended Exposure Limit (REL): an exposure limit recommended by NIOSH not to be exceeded.

—Short Term Exposure Limit (STEL): maximum concentration to which workers can be exposed for fifteen minutes for only four times throughout the day with at least one hour between exposures.

—Time Weighted Average (TWA): the average time, over a given work period of a person's exposure to a chemical or an agent, determined by sampling for the contaminant throughout the time period.

Various provisions of the act define general record-keeping and reporting procedures. Employers must maintain medical and chemical exposure records for at least 30 years. Employees must be permitted to see and copy their records upon written request. Specific standards may contain additional record-keeping and reporting requirements.

Emergency standards may be issued if required to protect workers from "grave dangers" posed by substances determined to be toxic or physically harmful or from new hazards. These standards are effective immediately upon publication, but are only valid for six months. Many final standards began as emergency standards.

An OSHA cancer policy, promulgated in 1980, defines criteria to be used to identify, classify, and regulate chemicals with potential human carcinogenic effects. The implementation of the policy was placed on hold by the Reagan administration, which issued a federal policy on procedures federal agencies should use when deciding whether or how to regulate carcinogens. The policy is still in effect but not in use.

Of greater current impact is the Hazard Communication Standard (HCS), which was issued in 1983 and took effect November 1985 for chemical firms and in May 1986 for certain manufacturing firms. The overall goal of the standard is to achieve implementation of risk management and safety programs by regulated employers. These programs are intended to inform employees of the hazards which may be encountered in the workplace and methods to minimize both the probability and severity of potential harm. Unlike other portions of the OSH Act, HCS regulations apply to federal employees as well as private sector employees.

The HCS as initially promulgated applies only to facilities in Standard Industrial Classification (SIC) codes 20 through 39. Acting under a federal appeals court order, OSHA issued a final rule August 1987 which expands the applicability of the HCS to all industries with more than ten workers, including construction. Newly covered employers must be in compliance with the standard by May 1988.

According to the standard, employers must furnish employees with instruction on the nature and effects of toxic substances with which they work, either in written form or in training programs. Instruction generally must include:

1. The chemical and common names of the substance;
2. The location of the substance in the workplace;
3. Proper and safe handling practices;
4. First aid treatment and antidotes in case of overexposure;
5. The adverse health effects of the substance;
6. Appropriate emergency procedures;
7. Proper procedures for cleanup of leaks or spills;
8. Potential for flammability, explosion, and reactivity; and
9. The rights of employees under this rule.

Most of this information is available from the Material Safety Data Sheet (MSDS), a document containing standardized information about the properties and hazards of listed toxic substances. Manufacturers, importers, and distributors of listed toxic substances are required to prepare and provide MSDSs to their direct purchasers.

A suggested MSDS form is shown in Figure 4.1. No items on an MSDS may be left blank. If information is not available or not applicable, the entry must be marked with "n/a." MSDSs must be in English, although an additional copy in another language may be used, if appropriate.

State-enacted Right-To-Know laws are related to the HCS, and these terms are often used interchangeably. However, Right-To-Know laws may be only similar to, or may differ substantially from, the HCS. States with approved OSHA state plans must establish standards equivalent to the HCS, although variations may be allowed if they are at least as stringent as the federal standard and do not interfere with interstate commerce. In other states, the HCS applies as promulgated by OSHA.

OSHA issued a Hazardous Waste Operations and Emergency Response interim final rule in December 1986, which is scheduled to become effective as a final rule in 1988. It is intended to protect workers at both uncontrolled and licensed waste sites, in addition to emergency response workers.

The current regulations will no doubt be altered before being adopted in final form, but the concerns addressed are likely to remain unchanged. In

Figure 4.1
Material Safety Data Sheet

Approved OSHA MSDS Form

Material Safety Data Sheet
May be used to comply with
OSHA's Hazard Communication Standard,
29 CFR 1910.1200. Standard must be
consulted for specific requirements.

U.S. Department of Labor
Occupational Safety and Health Administration
(Non-Mandatory Form)
Form Approved
OMB No. 1218-0072

IDENTITY (*As Used on Label and List*)

Note: Blank spaces are not permitted. If any item is not applicable, or no information is available, the space must be marked to indicate that.

Section I

Manufacturer's Name

Emergency Telephone Number

Address (*Number, Street, City, State, and ZIP Code*)

Telephone Number for Information

Date Prepared

Signature of Preparer (*optional*)

Section II — Hazardous Ingredients/Identity Information

Hazardous Components (Specific Chemical Identity, Common Name(s))	OSHA PEL	ACGIH TLV	Other Limits Recommended	% (optional)

Section III — Physical/Chemical Characteristics

Boiling Point		Specific Gravity (H_2O = 1)	
Vapor Pressure (mm Hg.)		Melting Point	
Vapor Density (AIR = 1)		Evaporation Rate (Butyl Acetate = 1)	

Solubility in Water

Appearance and Odor

Section IV — Fire and Explosion Hazard Data

Flash Point (Method Used)	Flammable Limits	LEL	UEL

Extinguishing Media

Special Fire Fighting Procedures

Unusual Fire and Explosion Hazards

(Reproduce locally) OSHA 174, Sept. 1985

Section V — Reactivity Data

Stability	Unstable		Conditions to Avoid
	Stable		

Incompatibility (*Materials to Avoid*)

Hazardous Decomposition or Byproducts

Hazardous Polymerization	May Occur		Conditions to Avoid
	Will Not Occur		

Section VI — Health Hazard Data

Route(s) of Entry: Inhalation? Skin? Ingestion?

Health Hazards (*Acute and Chronic*)

Carcinogenicity: NTP? IARC Monographs? OSHA Regulated?

Signs and Symptoms of Exposure

Medical Conditions
Generally Aggravated by Exposure

Emergency and First Aid Procedures

Section VII — Precautions for Safe Handling and Use

Steps to Be Taken in Case Material Is Released or Spilled

Waste Disposal Method

Precautions to Be Taken in Handling and Storing

Other Precautions

Section VIII — Control Measures

Respiratory Protection (*Specify Type*)

Ventilation	Local Exhaust	Special
	Mechanical (*General*)	Other

Protective Gloves | Eye Protection

Other Protective Clothing or Equipment

Work/Hygienic Practices

Page 2

* U.S.G.P.O. 1986-491-529/45775

its interim final form, the standards mandate site-specific safety and health programs; site evaluations and control programs; employee training; medical surveillance; engineering controls; air monitoring; safe work practices; and development of decontamination and emergency response procedures.

There are other portions of the OSH Act in addition to those previously discussed which pertain to hazardous materials management under specific circumstances. Subpart H contains regulations which specify safe handling and storage practices for compressed gases, flammable and combustible liquids, and explosives and blasting agents. Regulations in Subpart I address the application, use, and design of personal protective equipment. Various other OSH Act regulations deal with record-keeping, reporting, inspections, and miscellaneous requirements.

Enforcement

The OSH Act is enforced by the U.S. Department of Labor (DOL) through procedures that are specified in the act and administered by OSHA, which is a department within the DOL. OSHA itself is essentially an enforcement organization with a majority of its employees holding positions as inspectors. The act provides for citations and penalties authorizing fines of up to $10,000 per violation.

State laws may be more stringent than federal regulations. Also, state laws remain in effect where no federal standards exist. The regulations encourage states to develop and operate their own job safety and health programs, thereby becoming what is termed a "state plan state." Approximately half of the states had some type of approved OSH Act program as of December 1986.

Those states where concurrent federal jurisdiction of the OSH Act has been suspended, making the state the sole regulatory authority, are California, Michigan, Nevada, New Mexico, North Carolina, Oregon, South Carolina, Vermont, Virginia, and Washington. Status of the response to the HCS by states with an OSH Act plan is shown in Table 4.3.

Relationship to Other Regulations

The OSH Act overlaps with several occupation-specific regulations including the Toxic Substances Control Act (TSCA) and the Federal Insecticide, Fungicide, and Rodenticide Act (FIFRA). When a substance comes under regulation under TSCA, OSHA is affected if that substance appears in the workplace. In the case of pesticides, EPA is concerned with environmental effects, whereas OSHA is concerned with health effects on agricultural workers and pesticide applicators. OSHA and EPA regulate different

Table 4.3
Status of State Plan States' Response to Hazard Communication Standard

Right-To-Know Laws and Standards in State Plan States

States	States With Right To Know Laws	States Which Have Already Adopted Identical Standard	Different Standard
Alaska	x	-	x
Arizona	-	x	-
California	x	-	x
Connecticut	x	-	-
Hawaii	-	-	x
Indiana	-	x	-
Iowa	x	-	x
Kentucky	-	x	-
Maryland	x	-	x
Michigan	x	-	-
Minnesota	x	-	x
Nevada	-	x	-
New Mexico	-	x	-
New York	x	x	-
North Carolina	-	-	x
Oregon	x	-	x
South Carolina	-	x	-
Tennessee	x	x	-
Utah	-	x	-
Vermont	-	-	x
Virginia	-	x	-
Virgin Islands	-	x	-
Washington	x	-	x
Wyoming	-	-	x

aspects of other health-related issues such as asbestos, vinyl chloride, cancer policies, and hazard labeling.

The HCS does not apply to hazardous wastes regulated by EPA under the Resource Conservation and Recovery Act (RCRA). However, the recently issued Hazardous Waste Operations and Emergency Response regulations apply to workers cleaning up hazardous waste sites under the Superfund law or working at facilities permitted under RCRA.

Any labels required by OSHA may not conflict with the labeling requirements of the Hazardous Materials Transportation Act (HMTA). DOT promulgates rules that address the safety of transportation crews, maintenance personnel, and the traveling public, and consequently overlaps with similar responsibilities under the OSH Act.

Potential Future Developments

Unlike many environmental statutes, the original OSH Act has remained virtually unchanged, although new regulatory programs have been added since its inception. One of these additions, the cancer policy, has been essentially suspended.

Another, the HCS, is being implemented in spite of court challenges, and its complete judicial review process will probably take several years to complete. Many of the suits to date concern the appropriate treatment of information which industries claim is confidential business information. Upon extension of the HCS to include all industries, there has been significant support from the regulated community for a separate standard which would be applicable to construction industries. Supporters contend that the standard, as now written, is aimed mainly at chemical and manufacturing firms. Many construction managers feel that a separate standard could be drafted which would recognize the temporary nature of construction projects, the transient workforce usually involved, and prior training that workers may have received.

Other recent interim final rules applying to hazardous waste operations and emergency response will probably not result in significant litigation until after the final regulations have been issued. Two major areas appearing to require clarification are the scope of the regulations and the requirements for training, according to both industry and OSHA officials.

Various substances are continually coming under review for suspicion of causing cancer or other illnesses in humans. A leading candidate for more stringent regulation is currently formaldehyde, principally because of recent studies which have prompted EPA to classify it as a probable human carcinogen. It appears likely that the current exposure limit will be reduced at least by 50 percent.

OSHA has announced its intention to propose new workplace health standards for various other substances and to review existing standards where warranted. These standards will likely be shaped by resulting litigation, with organized labor generally attesting the standard as too permissive and industry protesting it as too stringent. In an effort to reduce the time and expense involved in such actions, OSHA has recently announced its intention to negotiate with interested parties before it sets new standards. Rather than unilaterally proposing a new rule and waiting for opposing parties to challenge it, OSHA hopes to achieve significant concensus and support from all interested groups before a standard is issued.

NATIONAL ENVIRONMENTAL POLICY ACT (NEPA)

NEPA defines a national environmental policy committed to use all practicable means to conduct federal activities in a way that will promote general welfare and be in harmony with the environment. It requires federal agencies to address the environmental consequences of their actions through preparation of an Environmental Impact Statement (EIS). The act also created the Council on Environmental Quality (CEQ), which issues standards to determine what material is required to be included in an EIS.

Background

NEPA was drafted to reflect a concern about the environmental consequences of the actions of federal agencies, and its principal objective was to encourage a productive and enjoyable harmony between people and their environment. The act defined the first environmental policy ever enacted by Congress.

NEPA does not directly address hazardous materials specifically and is relevant to the topic of hazardous materials management mainly because of its status as a historical precedent. For the most part, hazardous materials were not considered to be a public nuisance or an environmental health concern when NEPA was drafted in 1969. For example, the 1970 edition of the annual report by the CEQ did not even mention hazardous wastes. However, NEPA does represent a critical milestone in the general philosophy of environmental regulation and provides a foundation for virtually all environmental laws that followed.

NEPA was signed on January 1, 1970 (almost a year prior to the creation of EPA), and it was heralded by many to be the beginning of an environmentally oriented legislative decade. The act is often referred to as the Magna Carta or Bill of Rights of environmental legislation.

NEPA has spawned extensive legislation. In fact, some experts believe that it may have led to more lawsuits than all the other federal environmental laws combined. When violations of NEPA are alleged, any party that can show any evidence of resulting injury has standing to sue.

Ironically, the Supreme Court has found that NEPA creates no judicially enforceable substantive rights, but imposes only procedural requirements. However, the impact of those procedural requirements has been of such magnitude that virtually every federal agency has incorporated NEPA requirements into its routine procedures.

Further, the provisions in NEPA are subject to litigation, and the major cases have supported NEPA requirements. Therefore, it does not appear that the lack of regulatory power has hindered the effectiveness of the act in achieving its goal.

Content

NEPA is divided into two titles. The first title identifies environmental policy and goals and methods for accomplishing those goals. The second title creates the CEQ and defines its responsibility.

Probably the most important provision of NEPA is contained in section 102(2)(c). This section requires that a detailed statement (EIS) be prepared for every major federal action that significantly affects the quality of the human environment. The following must be addressed:

1. The environmental impact of the proposed action;
2. Any adverse environmental effects that cannot be avoided should the proposal be implemented;
3. Alternatives to the proposed action;
4. The relationship between local short-term activities and long-term enhancement of productivity of our environment; and
5. Any irreversible and irretrievable commitments of resources which would occur should the proposed action be implemented.

It is important to note that, unlike other regulations addressed in this book, NEPA does not prohibit any activities. Rather, it requires documented evaluation of the potential impacts of actions which come under its regulation.

NEPA applies to federal agencies only, and EISs are prepared by the responsible federal agency. States, local agencies, and private parties may assist or be required to assist. The analysis of the data and the conclusions must, however, be the work of the responsible federal agency.

The CEQ issues standards on what is required to be addressed in an EIS and monitors progress toward achieving national environmental goals. The CEQ is also responsible for conducting studies concerning broad areas such as policies, programs, standards, mediation, public involvement, and international cooperation. Since its inception, the council has carried out its responsibilities for a wide range of environmental concerns. Major activities have been undertaken in the areas of potentially hazardous substances, such as pesticides, toxic substances, hazardous waste, nuclear waste, and radiation. The council has sponsored research and published special reports on subjects such as integrated pest management, carcinogens in the environment, and hazardous waste.

In addition to NEPA's EIS requirements, the act contains seven other procedural action requirements, listed below. Federal agencies are required to

1. Utilize an interdisciplinary approach to planning and decision making;
2. Give appropriate consideration to unquantified environmental values;
3. Study and develop resolutions concerning conflicts over use of resources;
4. Recognize the worldwide and long-range character of environmental problems;
5. Aid in the dissemination of usable environmental information;
6. Provide ecological information for resource-oriented projects; and
7. Provide assistance to the CEQ.

These provisions are commonly considered essentially appendages to the EIS requirement. However, court decisions have supported at least some of these provisions as imposing duties which are both independent of and wider in scope than NEPA's EIS requirement. Whether these will come to play a more important role in the administration of NEPA law remains to be determined.

Enforcement

NEPA applies to federal agencies only, and the responsible federal agency must decide whether or not an EIS is required. The final draft of any EIS must be filed with EPA. EPA's Office of Federal Activities (OFA) has responsibility for: review and oversight of the environmental activities of other federal agencies as required by NEPA, review of all EISs written under NEPA, preparation of EISs on EPA programs, international environmental assessment, and performance of administrative support for the day-to-day operations of OFA and its regional counterparts.

Many states, following the lead of the federal government, have enacted NEPA-like laws. However, these do not serve in lieu of the federal NEPA program.

Relationship to Other Regulations

Some statutes have excluded certain activities of specific agencies from NEPA coverage. EPA is exempt from many of NEPA's requirements when engaged in regulatory activities. For example, EPA does not have to prepare an EIS prior to proposed environmental rule-making activities.

EPA is granted a role of "roving review" by section 309 of the Clean Air Act (CAA) which authorizes the agency to review and comment on the environmental impact of proposed legislation, regulations, or federal actions requiring an EIS. If any of these are determined to be unsatisfactory from the standpoint of public health or welfare, EPA must publish notice of its determination and refer the matter to the CEQ.

Many activities pertaining to hazardous materials management do not reference NEPA directly, but have been influenced by NEPA. For instance, a hazardous waste management facility generally is not required to produce an EIS, but many permitting requirements originated in NEPA-mandated regulations. For example, numerous requirements of a Part B Application under the Resource Conservation Recovery Act (RCRA) are similar to information required in an EIS.

Potential Future Developments

More than fifteen years after the passage of NEPA, the concept of an EIS has become relatively noncontroversial. However, any administration efforts to reduce NEPA's importance will probably increase the level of NEPA-related litigation.

In April 1986 a major relaxation of "worst case analysis" under NEPA was announced. The intention was to eliminate the need to develop scenarios for unlikely disasters and allow agencies to concentrate on the more likely problems that could arise from proposed projects.

The CEQ is expected to continue its efforts to conduct special studies in the area of potentially hazardous substances, particularly with emphasis toward improving independent analyses and coordination among federal agencies.

5

AIR AND WATER QUALITY REGULATIONS

A land ethic for tomorrow . . . should stress the oneness of our resources and the live-and-help-live logic of the great chain of life. If, in our haste to "progress," the economics of ecology are disregarded by citizens and policy makers alike, the result will be an ugly America.

—Steward Lee Udall,
The Quiet Crisis, 1963

With a national consciousness focused on environmental quality, major legislation was drafted in the early 1970s to address those areas where contamination is often dramatically visible and brings immediate concern: air and water. Resulting laws focused on point sources, which are discrete conveyances (such as pipes, ditches, conduits, and containers) from which pollutants are or may be discharged. The air and water laws addressed point sources which commonly became known as the "four pipes": factory smokestacks, automobile tailpipes, and city and factory sewers.

It became obvious that issues broader and less definitive than immediate health risks were involved. It was desirable to have clean water not only for drinking, but also for recreation; desirable to have clean air not only to protect public health, but also from an aesthetic viewpoint. Therefore, the definition of "primary" and "secondary" goals became concepts inherent in drafting many of these regulations and are issues still under debate today.

CLEAN AIR ACT (CAA)

The CAA was enacted to achieve air quality levels protective of public health and welfare, and it provides the regulatory framework for prevention and control of discharges into the air. The following are the overall goals of the act:

1. Protect and enhance the quality of the nation's air resources;
2. Implement a national research program to control air pollution;
3. Provide technical and financial assistance to state and local governments; and
4. Support development of regional air pollution control programs.

The act establishes primary and secondary standards for air quality, termed National Ambient Air Quality Standards, which are applicable to criteria pollutants. Hazardous pollutants are regulated by National Emissions Standards for Hazardous Air Pollutants. The act requires EPA to formulate plans to control pollutants through its own actions or through action by the states.

Background

The first CAA in the United States was passed in 1963 in response to air pollution brought about by urban and industrial development and the increasing use of motor vehicles. The act subsequently was amended in 1965 and 1967, with significant amendments also adopted in 1970 and 1977.

The 1970 amendments resulted in a totally restructured regulatory scheme which constitutes the basic CAA of today. Through these amendments, Congress authorized the establishment of stringent uniform national ambient air quality standards to be determined by the newly created Environmental Protection Agency. The 1977 amendments attempted to standardize the basis for rule making under the act and added provisions to guarantee that areas with clean air would not be allowed to deteriorate.

Content

The CAA requires EPA to do the following:

1. Identify air pollutants;
2. Set national air quality standards;
3. Formulate plans to control air pollutants;
4. Set standards for sources of air pollution; and
5. Set standards limiting the discharge of hazardous substances into the air.

The main sections of the CAA that deal with hazardous materials are ambient air quality standards, the regulation of hazardous air pollutants, and new pollution source performance standards. Unlike some other environmental regulations, which have virtually self-contained sections dealing with hazardous materials, these subject areas are interrelated and dependent on other compliance concepts. Three major, interrelated categories must be addressed for each emission source in order to determine the requirements for regulatory compliance. The categories are type of pollutant, location, and type of source.

The act addresses two major types of pollutants: criteria pollutants and hazardous pollutants. Criteria pollutants are regulated by primary standards designed to protect public health and by secondary standards designed to protect the public welfare. These standards are the National Ambient Air Quality Standards (NAAQS). Primary and secondary NAAQS have been promulgated for the following commonly detected air pollutants: sulfur oxides, particulate matter, nitrogen dioxide, carbon monoxide, photochemical oxidants (measured as ozone), and lead. Characteristics and standards for these air pollutants are given in Table 5.1. The standard for particulate matter was originally applicable to total particulate matter, but was revised in 1987 to address only particles smaller than 10 μm. The limitation of particulates from point sources provides for control of toxic and hazardous pollutants which enter the atmosphere as particles of a regulated size. The standards setting procedure for NAAQS is depicted in Figure 5.1.

Figure 5.1
Standards Setting Procedure for NAAQS

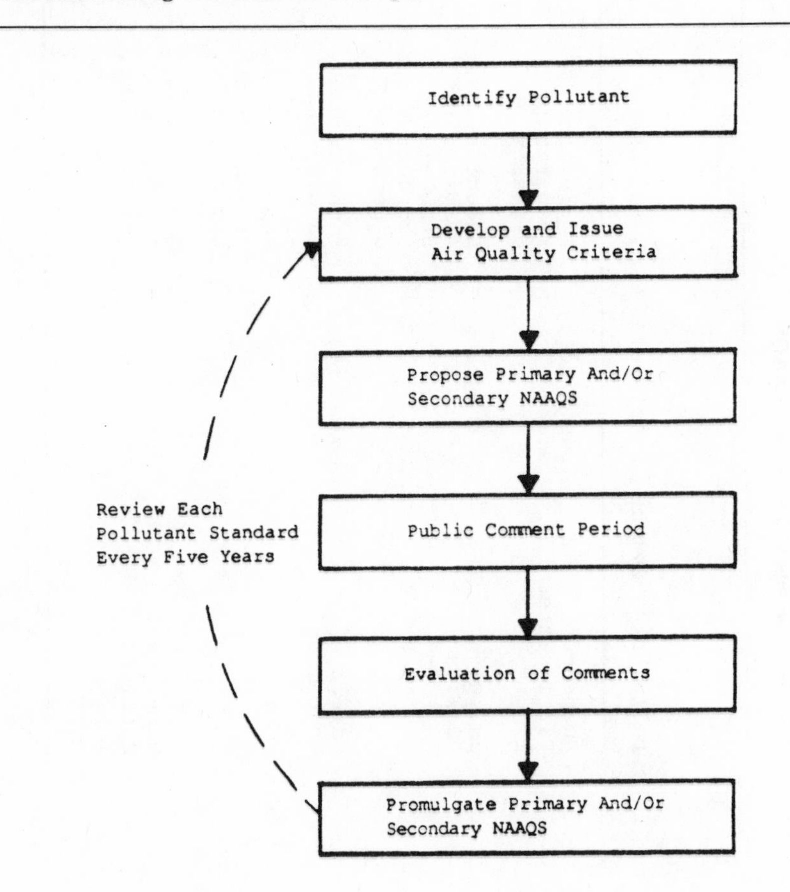

Table 5.1
Criteria Pollutants Regulated by NAAQS

Pollutant	Symbol	Description	Health Effects	Other Effects	Primary Standard ($\mu g/m^3$)	Secondary Standard ($\mu g/m^3$)
Suspended Particulates 10 um	SP	Dust, soot, smoke	Aggravates respiratory diseases	Aesthetic concerns	150[1] 50[2]	150[1] 50[2]
Sulfur Dioxide	SO_2	Corrosive	Associated with respiratory problems, aggravates heart disorders	Can cause acid rain	80[2] 365[1] --	-- -- 1,300[3]
Carbon Monoxide	CO	Colorless, odorless	Fatal at high levels, can cause nausea, headaches, dizziness, breathing difficulty	Impairs judgment and perception	10,000[4] 40,000[5]	10,000[4] 40,000[5]
Nitrogen Dioxide	NO_x	Highly reactive	Fatal at high concentrations	Can react with hydrocarbons to form ozone	100[2]	100[2]

Pollutant	Symbol	Description	Health Effects	Other Effects	Primary Standard ($\mu g/m^3$)	Secondary Standard ($\mu g/m^3$)
Ozone	O_3	Pungent smelling, faintly bluish	Irritates mucous membranes	Principal constituent of smog	235[5]	235[5]
Lead	Pb	Mainly from auto emissions	Affects blood forming, nervous and kidney systems	Young children are at high risk	1.5[6]	1.5[6]

[1] 24-hour maximum

[2] Annual arithmetic mean

[3] 3-hour maximum

[4] 8-hour maximum

[5] 1-hour maximum

[6] Calendar quarter

Hazardous pollutants are regulated by the National Emission Standards for Hazardous Air Pollutants (NESHAPs), and the following substances have been listed as hazardous air pollutants: asbestos, beryllium, mercury, vinyl chloride, benzene, radionuclides, inorganic arsenic, and coke oven emissions. Hazardous pollutants differ from criteria pollutants in that they are usually source localized and can be technically difficult and costly to control.

The NESHAPs are designed to control the source of the hazardous pollutant, but do not regulate the levels of these substances in the ambient air. NESHAPs also do not regulate by substance identification alone, but are coupled with a source because not all sources of the regulated pollutant are subject to NESHAPs limitations. For example, the emission of mercury and radionuclides from coal-fires power plants are not regulated even though an emission standard has been established for both pollutants. Compliance requirements include submitting a compliance status information form, notifying the appropriate agency, and performing emissions testing, monitoring, and reporting. Pollutants and related sources regulated by NESHAPs are shown in Table 5.2

The standards setting procedure for NESHAPs is depicted in Figure 5.2. This procedure, in contrast with the procedure for establishing NAAQS, requires an initial health assessment before a standard can be promulgated. EPA maintains lists of pollutants it intends to list as hazardous, of pollutants evaluated but not listed, and of pollutants under review. EPA has stated its intent to regulate the following substances under NESHAPs: carbon tetrachloride, ethylene dichloride, perchlorethylene, cadmium, ethylene oxide, chloroform, 1-3-butadiene, and chromium.

Source location determines the enforcement area and the regulatory agency in authority. There are about 240 discrete enforcement areas, known as Air Quality Control Regions (AQCRs). There are interstate or intrastate areas with common meteorological, industrial, and socioeconomic factors. Each area is considered as a single unit for some purposes of air pollution control.

The 1977 amendments introduced two new location-related concepts, which in effect reduced the importance of AQCRs by requiring the designation of areas which meet or exceed NAAQS. Areas in which these standards are met are known as attainment areas and are defined as one of the classes listed below:

Class Description:

I Virtually no emission increase allowed

II Modest emission increases allowed

III Greater emission increases allowed than allowed for Class II

Table 5.2
Pollutants and Sources Regulated under NESHAPs

Pollutant	Sources Regulated
	Promulgated Standards
Radon-222	Underground uranium mines
Beryllium	Extraction, ceramic, and propellant plants; foundries; and incinerators which process beryllium
Beryllium	Rocket motor test sites
Mercury	Mercury ore processing; mercury chlor-alkali cell production; sewage treatment plant sludge incinerators/dryers
Vinyl Chloride	Ethylene dichloride, vinyl chloride, and polyvinyl chloride production processes
Radionuclides	Department of Energy facilities
Radionuclides	Facilities licensed by Nuclear Regulatory Commission or federal facilities not covered under Subpart H
Benzene	Benzene service equipment (equipment leaks)
Radionuclides	Elemental phosphorus plants - calciners and nodulizing kilns
Asbestos	Asbestos mills; roadways; asbestos manufacturing processes; demolition and renovation; spraying; fabricating; insulating materials; waste disposal
Volatile Hazardous Air Pollutants*	Volatile hazardous air pollutant service equipment (equipment leaks)

* For which a standard for equipment leaks of the substance has been promulgated. Currently applicable to benzene.

March, 1986

These areas are regulated by Prevention of Significant Deterioration (PSD) regulations, which are designed to prevent deterioration of existing air quality.

Figure 5.2
Standards Setting Procedure for NESHAPs

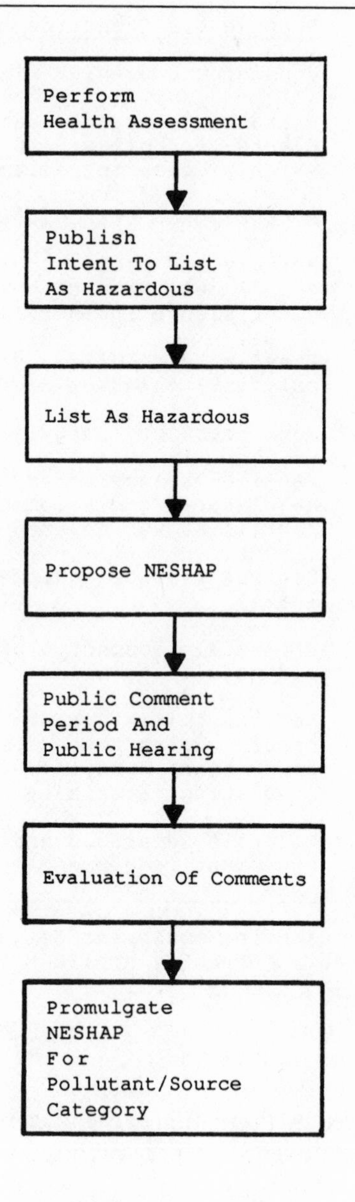

Areas in which NAAQS are exceeded are known as nonattainment areas (NA). These areas are governed by regulations, which are defined on a case-by-case basis, and are intended to improve air quality while allowing for some growth. Generally, existing sources must apply reasonably available control technology (RACT) and new sources must meet the lowest achievable emission rate (LEAR).

Type of source involves the age of the facility, mobility, and type of industrial process involved. Whether a source is new or existing is defined by dates specified in the regulations. Modification of an existing facility can make it subject to regulations governing new facilities.

Mobile sources are vehicular. A summary of regulatory requirements for motor vehicles is presented in Table 5.3. Other emission standards apply to aircraft and can only be issued if the Secretary of Transportation determines that they do not create safety problems.

Table 5.3
Summary of CAA Motor Vehicle Regulations

Emission testing and certification:

- Domestic vehicles and engines
- Imported vehicles and engines

New car emission regulations:

- Hydrocarbons
- Carbon monoxide
- Nitrogen oxides
- Particulate matter
- Smoke

Research areas:

- Total ban on lead in gasoline
- Diesel fuel quality
- Health effects of fuel and fuel additives
- NOx emission reduction to achieve 0.41 grams/mile

Stationary sources are regulated by technology-based guidelines, which depend partially on the location of the emission source. The control categories which are applicable to stationary sources are the following:

—New Source Performance Standards: applicable to newly constructed facilities potentially emitting 100 tons/year.

—Control Technique Guidelines: applicable to existing plants in areas that meet NAAQS.

—Reasonably Available Control Technology: applicable to existing plants where NAAQS are not met.

—Lowest Achievable Emission Rates: applicable to nonattainment areas.

—Best Available Control Technology (BACT): designed for use in PSD areas.

For new or modified stationary sources, the Standard Industrial Classification (SIC) determines whether a facility is subject to New Source Performance Standards (NSPS). NSPS are established for particular pollutants in certain industrial categories and are based on the best demonstrated system of continuous emissions reduction for that industrial category. These standards apply nationwide and are intended to discourage plants from moving to states with less stringent air pollution rules. The standards setting process for NSPS is shown in Figure 5.3.

A pollutant regulated under NAAQS may also be regulated under NSPS, but not all criteria pollutants are regulated by NSPS, and NSPS regulates pollutants that are not regulated by NAAQS. A tabular summary of pollutants regulated by the CAA and the related standards is shown in Table 5.4.

The imminent hazard provision of the CAA is found in section 303(a). It grants EPA broad authority to bring action in U.S. District Court to prevent or abate the emergency or hazard. In addition to bringing suit immediately to restrain the alleged pollution, EPA may issue orders to protect the health of threatened persons after consulting with state and local authorities.

Enforcement

Although a major thrust of amending the original CAA was to make compliance a federal rather than local goal, provisions in the act specify that state and local governments have primary responsibility for limiting air pollution, with guidance and financial assistance to be provided by the federal government.

The CAA regulations are to be implemented through State Implementation Plans (SIPs). The act provides considerable flexibility for states to determine how best to meet its requirements. Where a state does not have authority, EPA enforces the CAA through its regional offices. Permit requirements vary among states and may vary from region to region even when administered by EPA. Figure 5.4 shows the status of various states' PSD programs.

The Department of Transportation (DOT) enforces the CAA standards that are applicable to aircraft.

Relationship to Other Regulations

The CAA must be interpreted in conjunction with the Resource Conservation and Recovery Act (RCRA) where the incineration of hazard-

Figure 5.3
Standards Setting Procedure for NSPS

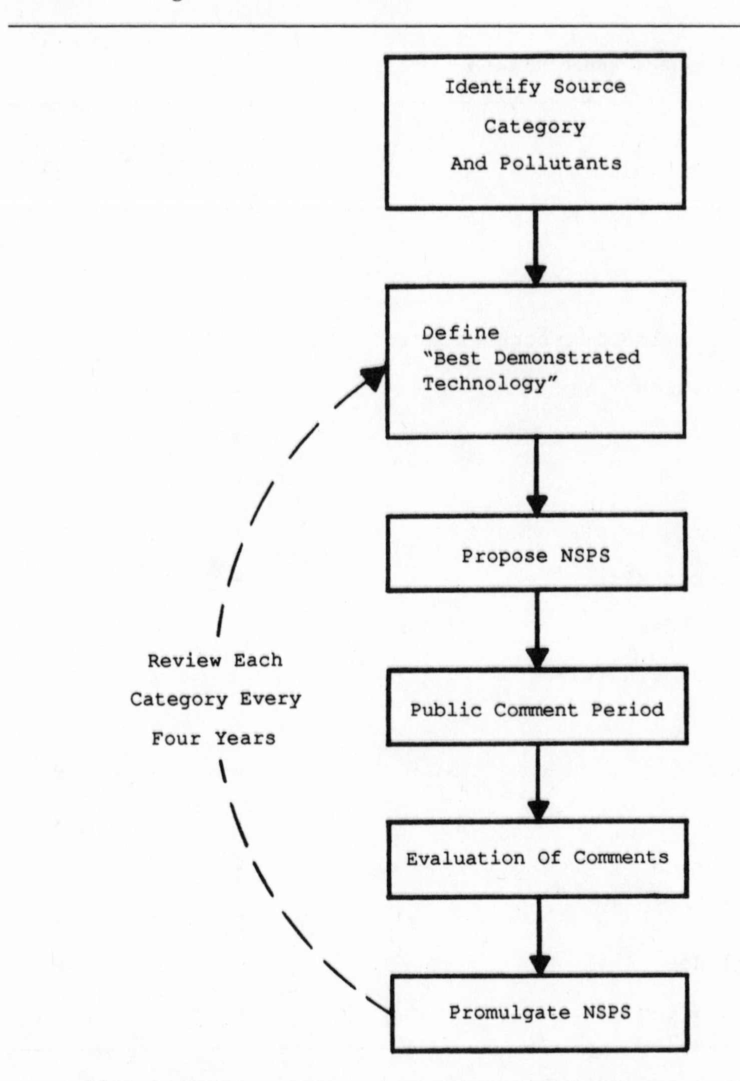

ous waste is involved. RCRA also regulates the emission of particulates at permitted hazardous waste piles, landfills, and land treatment operations. Radionuclides are regulated under section 112 of the act. The act bears a similarity to the Clean Water Act (CWA) because both use technology-based guidelines. Pesticide emissions regulated by the CAA may also be regulated by the Federal Insecticide, Fungicide, and Rodenticide Act (FIFRA). DOT has issued regulations for the purpose of enforcing CAA standards which are applicable to aircraft.

Table 5.4
Pollutants Regulated under the CAA

Pollutant	Standard		
	NAAQS	NSPS	NESHAP
Particulate matter	x	x	-
Sulfur dioxide	x	x	-
Nitrogen dioxide (nitrogen oxides)	x	x	-
Carbon monoxide	x	x	-
Ozone	x	-	-
Lead	x	-	-
Volatile organic compounds*	-	x	-
Sulfuric acid mist	-	x	-
Total fluorides	-	x	-
Total reduced sulfur**	-	x	-
Asbestos	-	-	x
Beryllium	-	-	x
Mercury	-	-	x
Vinyl chloride	-	-	x
Benzene	-	-	x
Radionuclides	-	-	x
Inorganic arsenic	-	-	x

* Generally excludes non-reactive volatile organic compounds

** Includes hydrogen sulfide, methyl mercaptan, dimethyl sulfide and dimethyl disulfide

Notes: NAAQS = National Ambient Air Quality Standards
NSPS = New Source Performance Standards
NESHAP = National Emission Standards for Hazardous Air Pollutants

Figure 5.4
State PSD Program Status

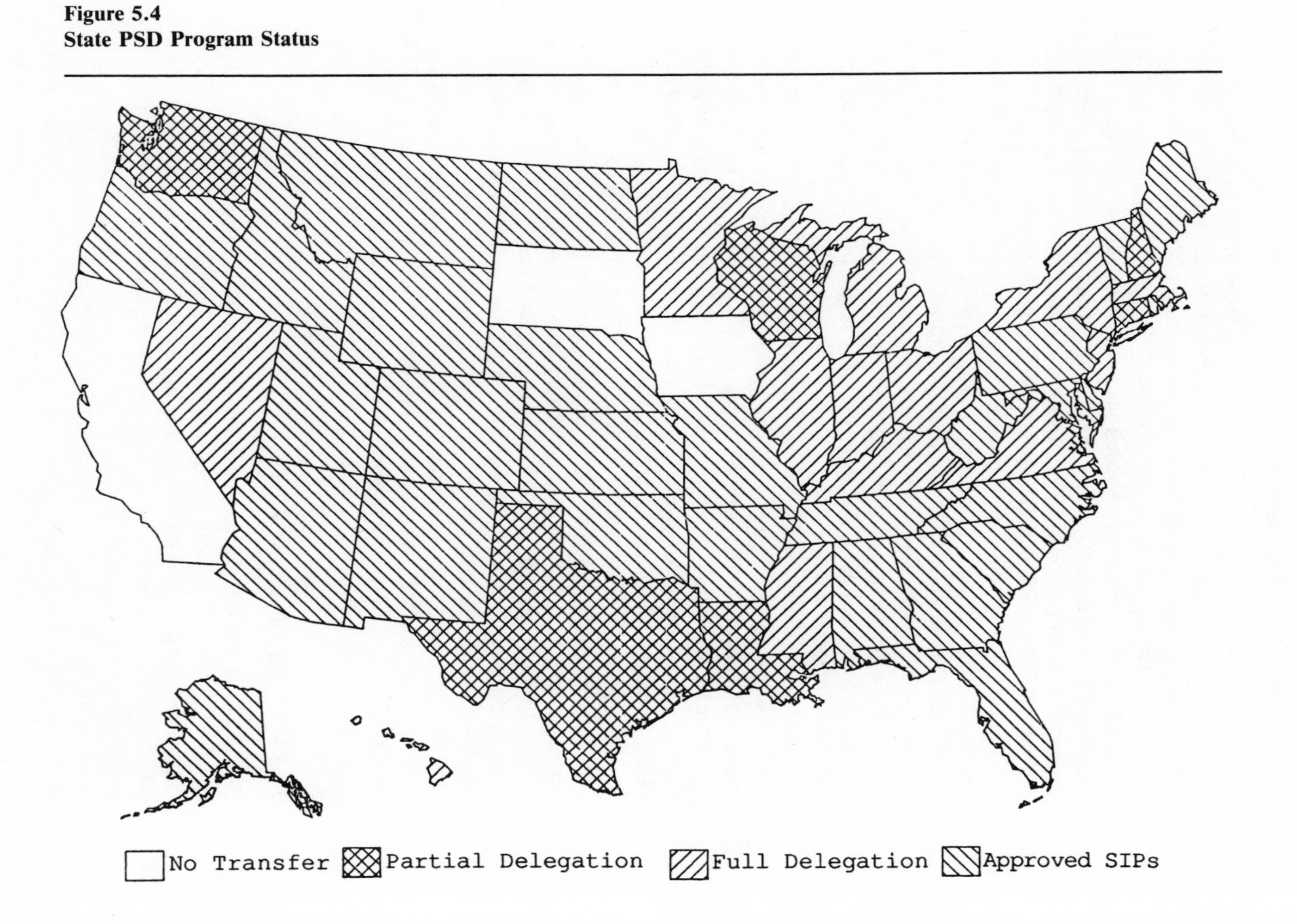

Potential Future Developments

A more stringent air pollution bill was introduced in the House in 1985, at least partially in response to the Bhopal chemical leak which claimed over 2,000 lives in India. Among other measures, the bill would have required EPA to list hazardous threshold levels for 85 chemical substances within a two-year period. EPA expressed serious reservations about the bill, while at the same time announcing a new toxic air pollutant strategy that turns over certain monitoring responsibilities to state and local government. Efforts to amend the CAA were not successful in either 1985 or 1986.

Congress is expected to continue its attempts to amend the CAA. Issues expected to be addressed include control of pollution that is thought to cause acid rain and the relationship of atmospheric pollution to thinning of the protective ozone layer. Because of the controversy and far-reaching economic implications surrounding both issues, obtaining the consensus necessary to amend the CAA will probably require lengthy and intensive legislative efforts.

The CAA set December 31, 1987 as the deadline for attainment of NAAQS for ozone and carbon monoxide. Throughout 1987, fourteen metropolitan areas and various other outlying areas were identified as being in non-compliance, with little or no chance of meeting the December deadline. Both sanctions and legislative efforts to extend the deadline are likely in an effort to remedy the problem. Some legislators favor extending the deadline to 1992. On the enforcement side, it is likely that EPA will initially utilize some of the milder sanctions that it is authorized to impose. Theoretically, sanctions which could be applied to the nonattainment areas include loss of federal highway and sewer aid and bans on certain construction.

Revising the standards for particulate matter in 1987 to include only particles smaller than 10 μm enabled various nonattainment areas for this standard to come into compliance. In an effort to remedy potential detrimental environmental impacts resulting from revising the standard, EPA has announced its intention to formulate new standards to protect visibility.

CLEAN WATER ACT (CWA)

The CWA controls discharges of effluent from point sources into waters of the United States. Both industrial and municipal discharges are considered point source pollution. The act establishes national technology-based effluent standards, and all point source discharges are required to comply with these standards.

Recognizing that accidental spills and other non-point source discharges are responsible for a large percentage of the pollutants introduced into national waters, the act also provides a number of mechanisms intended to deal with such discharges.

Background

Federal regulation of water quality dates to the Refuse Act of 1899, which was designed to protect navigation. However, it additionally prohibited all discharges into navigable waters or tributaries thereof unless a permit was obtained in advance from the U.S. Army Corps of Engineers (COE). This provision established a major milestone in federal efforts to regulate industrial aqueous discharges.

The present CWA is the successor to several water pollution control statutes that were passed in the 1950s and 1960s. For the most part, these precursor laws were not effective in controlling water pollution.

In 1971, Congress formulated a new approach to pollution control in amendments to the Federal Water Pollution Control Act (FWPCA), which subsequently became the CWA. These amendments, adopted in 1972, established the federal government, and specifically EPA, as the governmental body with primary responsibility for administering the national water pollution control program. The goal of the regulation is the elimination of all discharges into surface waters and the attainment of swimmable and fishable waters. Congress reauthorized an amended CWA in 1987 over a presidential veto.

Content

The CWA regulates various types of pollutants involved in specific types of discharges. The pollutant types addressed by the act are conventional, nonconventional, and toxic.

Conventional pollutants are substances that deplete oxygen in receiving waters, alter pH levels, or add suspended solids, fecal matter, oil, or grease. Nonconventional pollutants are neither conventional or toxic. Nonconventional pollutants can be reclassified as conventional or toxic at a later date, if warranted by subsequent evaluation. Current examples of nonconventional pollutants include ammonia and phosphorus.

Hazardous materials are usually regulated under the CWA as toxic pollutants. These pollutants, widely known as priority pollutants, originally consisted of 65 toxic chemicals. Since these included several generic classes of chemicals, EPA developed a list of 129 specific chemicals to more accurately define which substances are regulated. Three were subsequently deleted, yielding a total of 126. Due to the original format of listing, the last chemical is still numbered 129. However, no chemical is currently assigned to numbers 17, 49, or 50. Priority pollutants are listed in Table A.1 in the appendix.

The CWA regulations include standards for the following types of pollution discharges: direct discharges, indirect discharges, sources that spill oil or hazardous substances, discharges of dredged or filled material, and the disposal of sewage treatment sludge.

It is convenient to associate four of these major provisions of the CWA with a key acronym, as follows:

Direct Discharges: NPDES
Indirect Discharges: POTW
Oil or Hazardous Substances: SPCC
Dredge and Fill: COE

These sections will be defined briefly, followed by a more detailed discussion of the provisions in them that relate to hazardous materials management.

Facilities that are direct dischargers into navigable waters must obtain a National Pollutant Discharge Elimination System (NPDES) permit. Requirements contained in NPDES permits are based on the source of effluent, national technology-based guidelines, and state water quality standards. The following types of controls may be applicable to direct dischargers: Best Practicable Technology (BPT), Best Conventional Technology (BCT), Best Available Technology (BAT), Best Engineering Judgment (BEJ), and Best Available Demonstrated Control Technology (BADCT).

Discharges into municipal sewers are considered to be indirect discharges and do not require a NPDES permit. Instead, industrial dischargers which discharge into publicly owned treatment works (POTWs) must comply with pretreatment standards. General standards address fire or explosion hazards, corrosive discharges, viscous obstruction, slug discharges, and heat sufficient to inhibit biological activities. Specific pretreatment standards are set by industrial categories, as described by the Standard Industrial Classification (SIC). These address the introduction of pollutants that are not susceptible to treatment, interfere with the operation of the treatment plant, or prevent sludge use or disposal. Pretreatment standards are generally intended to achieve BAT removal efficiencies and define effluent quality which must be obtained prior to discharge to a POTW.

Section 311 of the act prohibits the discharge of hazardous chemicals and oil in quantities as may be harmful into waters of the United States. It also requires certain industries to prepare Spill Prevention Control and Countermeasure (SPCC) plans to control oil pollution.

The CWA regulates the discharge of dredged or fill material into waters of the United States. The program includes all navigable and interstate waters and tributaries of both, all other waters whose use might affect interstate commerce, and wetlands adjacent to any of these waters. The program excludes ordinary agriculture; silviculture; and the maintenance of dams, levees, farm ponds, and forestry roads. The Army Corps of Engineers (COE) is the agency with primary authority for this portion of the act.

Major portions of the act that deal with hazardous materials are contained in the first three of the four provisions just discussed, and consist of water quality criteria, water quality standards, effluent limitations, effluent guidelines, and control of oil and hazardous substances discharges. For the most part, dredge and fill regulations do not address hazardous substances.

Water quality criteria, set by EPA, describe the level of specific pollutants that ambient water can contain and still be acceptable for one of the following specific use categories:

Class A—Water contact recreation, including swimming

Class B—Able to support fish and wildlife

Class C—Public water supply

Class D—Agricultural and industrial use

Water quality standards are set by the states, subject to approval by EPA. These standards must define the conditions required to maintain the intended use water quality. Per provisions of the CWA, existing uses of a body of water must be maintained (i.e., uses that degrade water quality resulting in a downgraded specific use category are not allowed).

Effluent guidelines define uniform national guidelines for specific pollutant discharges for each type of industry regulated. These are not, in fact, "guidelines," but are regulatory requirements. Effluent limitations are specific control requirements applicable to a specific point source discharge. Effluent limitations are based on consideration of both national effluent guidelines and state water quality standards.

The CWA regulates many toxic pollutants through the NPDES permitting system; and the relationship of water quality criteria, water quality standards, effluent guidelines, effluent limitations, and NPDES permit conditions is depicted in Figure 5.5.

Section 311 of CWA regulates the discharge of oil and hazardous substances where the discharge is not associated with an NPDES permit. This section identifies approximately 300 substances as hazardous when spilled or discharged and establishes the minimum quantity that, when spilled, must be reported to the National Response Center (NRC). This minimum quantity is known as the reportable quantity (RQ) and, depending on the nature of the substance, is defined as 1, 10, 100, 1,000 or 5,000 pounds. A list of RQs is shown in Table A.2 in the appendix.

The CWA requires formulation of a plan to minimize damage from hazardous substance discharges. This plan is known as the National Contingency Plan (NCP). The contents of this plan, published in 40 CFR 300, include:

1. Assignment of duties among federal agencies;
2. Identification, procurement, maintenance, and storage of equipment and supplies;
3. Designation of a strike force trained, prepared, and available to carry out the plan;
4. A system of surveillance to insure earliest possible notice of discharges;

5. Establishment of a national center to provide coordination and direction for operations in carrying out the plan;
6. Publication of a schedule identifying dispersants and chemicals, water in which they may be used, and safe quantities for use; and
7. Establishment of a system whereby states may act to remove such discharges and be reimbursed by the federal government.

Figure 5.5
Relationship of Elements Used in Defining NPDES Permit Conditions

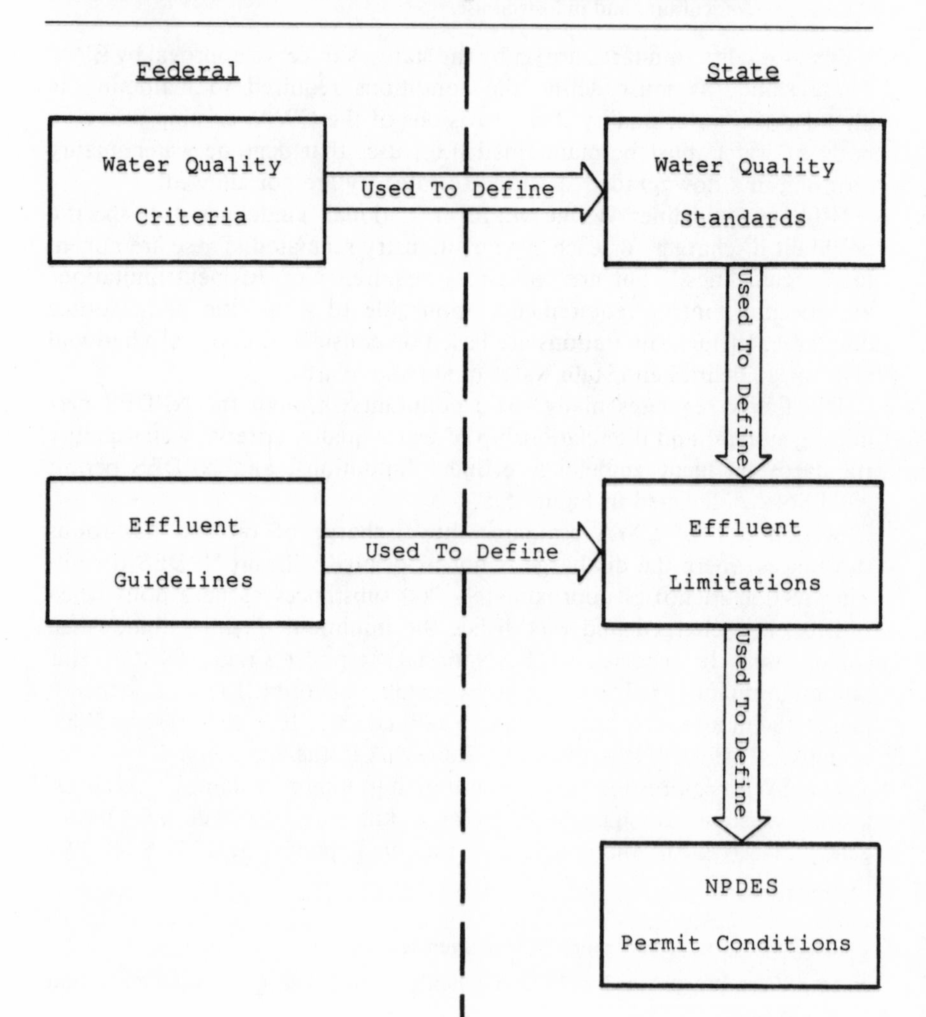

Initially, the regulations contained in section 405 of the CWA, Disposal of Sewage Sludge, had little relevance to the control of hazardous materials. However, the situation has changed, mainly due to a concern that pollutants released by indirect dischargers will concentrate in the sewage sludge produced by POTWs, thereby posing a threat to public health or the environment. Driven by 1987 amendments to the CWA, EPA has issued regulations to control toxic pollutants in sewage sludge. The approach involves numerical limitations for the most common pollutants, with regulations to address additional pollutants of concern to follow. Reviews to identify additional pollutants of concern are scheduled to occur every two years.

Section 504 of the CWA contains an imminent hazard provision allowing EPA to require cleanup of sites creating imminent and substantial endangerment to the public health or the environment. The section is applicable to control of point sources that discharge pollutants into surface waters.

In early 1987, a new Clean Water Act was passed over a presidential veto, via legislation termed the Water Quality Act of 1987 (WQA). The most controversial issue involved in the WQA was the funding level of the bill, which provides $18 billion in grants and loans over a nine-year period to help communities build new sewage treatment facilities. Important provisions relating to hazardous materials management include a program for controlling contaminated runoff from farms and urban areas, programs to clean up national lakes and estuaries, more stringent compliance requirements, and provisions to prevent less stringent clean-up standards. The compliance date for meeting clean-up standards for all types of pollutants has been designated as March 31, 1989. Also mandated were new deadlines for dealing with toxic contaminants in sewage treatment sludge.

Enforcement

States administer the act upon authorization by EPA, and states may set standards more stringent than those of the federal government. A list of states with various permitting authorities is shown in Table 5.5. The dredge and fill portion of the act (section 404) is administered by the COE.

Fines up to $10,000 and/or imprisonment can be levied against individuals or parties responsible for an illegal discharge of oil or other listed hazardous material, in addition to civil penalties of up to $5,000 for each discharge. Civil action against an owner, operator, or person in charge of the discharging facility can result in fines up to $250,000 if negligence or willful misconduct is proven. Section 311 authorizes the federal government to pass the costs of control and cleanup to the responsible party or parties.

Relationship to Other Regulations

Section 307(a) may be used to set effluent limitations for some toxic pollutants which are also regulated by other legislation. The CWA may

Table 5.5
CWA State Program Status

State	Approved State NPDES Permit Programs	Approved To Regulate Federal Facilities	Approved State Pretreatment Program
Arkansas	X	X	X
Alabama	X	X	X
California	X	X	-
Colorado	X	-	-
Connecticut	X	-	X
Delaware	X	-	-
Georgia	X	X	X
Hawaii	X	X	X
Illinois	X	X	-
Indiana	X	X	-
Iowa	X	X	X
Kansas	X	X	-
Kentucky	X	X	X
Maryland	X	-	X
Michigan	X	X	X
Minnesota	X	X	X
Mississippi	X	X	X
Missouri	X	X	X
Montana	X	X	-
Nebraska	X	X	X
Nevada	X	X	-
New Jersey	X	X	X
New York	X	X	-
North Carolina	X	X	X
North Dakota	X	-	-
Ohio	X	X	X
Oregon	X	X	X
Pennsylvania	X	X	-
Rhode Island	X	X	X
South Carolina	X	X	X
Tennessee	X	-	X
Vermont	X	-	X
Virgin Islands	X	-	-
Virginia	X	X	X
Washington	X	-	-
West Virginia	X	X	X
Wisconsin	X	X	X
Wyoming	X	X	-

define reportable quantities of materials requiring spill notification under the Comprehensive Environmental Response, Compensation, and Liability Act (CERCLA), and the CWA provision for drawing up the NCP was revised and incorporated into CERCLA. The CWA is similar to the Clean Air Act (CAA) in that it utilizes technology-based guidelines that the states must meet. All activities under the CWA except POTW construction and new source discharge permitting are specifically exempted from the provisions of the National Environmental Policy Act (NEPA).

Potential Future Developments

With the recent passage of a new Clean Water Act, no major legislative developments are anticipated in the near future. Instead, there will be significant activity on the part of regulatory agencies in defining the newly mandated programs and promulgating related regulations which are anticipated overall to result in more stringent regulation of toxic pollutants. There will also be significant activity in building new facilities which will have to meet more stringent standards. The dispersment of the CWA funding among the states is shown in Table A.3 in the appendix.

Certain provisions in the WQA support the move from technology-based standards to water quality-based standards. The latter are more difficult to implement because they vary with the characteristics of the receiving water. However, water quality-based standards have the theoretical advantage of assuring that the quality of the receiving water is adequate for its intended use.

In a situation similar to that created by CAA compliance deadlines, an estimated 800 wastewater treatment systems are in danger of being in noncompliance with the CWA effective July 1988, a deadline specified by the act. There are currently no legislative proposals to extend the deadline or ease regulatory requirements. There are indications, however, that EPA may be lenient toward selected systems that have at least begun to take measures to achieve compliance.

SAFE DRINKING WATER ACT (SDWA)

The overall intent of the SDWA is to ensure the quality of drinking water consumed by the public and to protect underground sources of water. The major provisions of the act establish two broad programs related to hazardous materials management: one regulating public water supply systems and the other regulating underground injection of wastes. The act also authorizes EPA to provide financial and technical assistance to state and local governments in research and study efforts. The EPA hotline number for information on the SDWA is 800-426-4791.

Background

Passage of the Clean Water Act (CWA) provided for the cleanup and protection of surface waters, but did not address the regulation of ground water, which is the source of about 50 percent of the U.S. water supplies. The intent of the SDWA was to fill this regulatory gap to protect underground aquifers, and to establish regulatory control over all public drinking water supply systems to ensure drinking water quality.

By 1970, an effort had started in the federal government to review and revise the 1962 Drinking Water Standards. In 1971, EPA asked a coalition of the National Academy of Sciences and the National Academy of Engineering to revise the 1968 Water Quality Criteria. Resulting investigations demonstrated that public confidence in the safety of drinking water supplies might in many instances be misplaced.

The SDWA addressed this concern by allowing EPA to establish federal standards to control the levels of harmful contaminants in drinking water supplied by public water systems. In drafting the law, Congress also provided EPA with regulatory control of underground injection of substances into wells which might contaminate ground water sources. The act was signed by President Ford on December 16, 1974. Significant amendments to the act were passed in 1977 and 1986.

Content

The major provisions of the act as passed in 1974 included:

1. The establishment of primary regulations for the protection of the public health;
2. The establishment of secondary regulations that are related to taste, odor, and appearance of drinking water;
3. The establishment of regulations to protect underground drinking water sources by the control of subsurface injection;
4. The initiation of research on health, economic, and technological problems related to drinking water supplies;
5. The initiation of a survey of rural water supplies; and
6. The allocation of funds to states for improving their drinking water programs through technical assistance, training of personnel, and grant support.

Provisions of the act that directly address the regulation of hazardous materials are found in section 1412 (drinking water standards) and section 1421 (underground injection).

For public water supply systems, the act defines primary drinking water regulations, which consist of standards for maximum concentration of contaminants, and standards for monitoring, analysis, reporting, and notification of users. Public systems are defined as those that provide piped water for human consumption to a minimum of fifteen service connections or 25 individuals daily. Standards vary for public systems depending on whether they are classified as community or noncommunity suppliers.

Among the primary standards applicable to public water supply systems are Maximum Contaminant Levels (MCLs) for fluoride, arsenic, mercury, lead, nitrates, radioactivity, various pesticides, and other inorganic and organic chemicals. A summary of current primary MCLs are presented in Table 5.6. These are intended to reflect the contaminant level that EPA believes is feasible to obtain considering the status of treatment technologies and related costs.

Table 5.6
Primary Drinking Water Standards

Contaminant	Maximum Contaminant Level (MCL) (mg/l unless otherwise indicated)
Arsenic	0.05
Barium	1
Cadmium	0.010
Chromium	0.05
Lead	0.05
Mercury	0.002
Selenium	0.01
Silver	0.05
Nitrate (as N)	10
Fluoride	4
Endrin	0.0002
Lindane	0.004
Methoxychlor	0.1
Toxaphene	0.005
2,4-D	0.1
2,4,5-TP Silvex	0.01
Trihalomethanes (Total)	0.10
Turbidity	1-5*
Radium-226 & 228	5**
Gross Alpha	15**
Tritium	20,000**
Strontium-90	8**
Coliform Bacteria	1+
Benzene	5++
Vinyl Chloride	2++
Carbon Tetrachloride	5++
1,2-Dichloroethane	5++
Trichloroethylene (TCE)	5++
P-Dichlorobenzene	75++
1,1-Dichloroethylene (DCE)	7++
1,1,1-Trichloroethane (TCA)	200++

* Turbidity Units. The general standard is one, but up to five may be allowed with a special exemption.

** pCi/L

+ Per 100 ml

++ ug/l

For the most part, secondary MCLs address aesthetic concerns and, as such, do not normally regulate hazardous materials. Secondary MCLs are advisory only, and cannot be enforced unless required by state or local regulations. Because of their advisory status, in some cases secondary MCLs are more stringent than the primary, or enforceable, MCLs. In essence, this allows EPA to recommend a higher water quality than that quality mandated by enforcement.

In addition to establishing MCLs, EPA was required to establish Recommended Maximum Contaminant Levels (RMCLs) for contaminants that may cause adverse health effects. These are required to be set at a level at which no known or anticipated adverse effects on health occur and are to include a safety margin. The 1986 amendments have redefined RMCLs as Maximum Contaminant Level Goals (MCLGs).

In the future, MCLs and MCLGs must be proposed and promulgated simultaneously. The MCL, based on the concentration which can be produced by use of best available treatment (BAT), is to be as close as is feasible to the MCLG, but must also take cost into consideration.

The 1986 SDWA amendments require EPA to set drinking water standards for 83 additional specific contaminants within three years; contaminants are listed in Table A.4. Of these 83 contaminants, only six are microbiological or turbidity components. The remaining 77 are potentially hazardous materials from the following categories: volatile organic compounds (VOCs), synthetic organic chemicals (SOCs), inorganic chemicals (IOCs) and radionuclides. The list also includes all substances which already have MCLs promulgated, to emphasize that EPA is charged with re-evaluating these standards, in addition to issuing new ones.

RMCLs (or MCLGs according to the new terminology) had already been promulgated for certain VOCs on the list, and these substances were among the first to have MCLs promulgated. The last eight standards listed in Table 5.6 were issued by EPA in July 1987, with compliance required effective January 1989.

For unregulated drinking water contaminants that have been detected in potable water supplies, EPA has issued health effect advisories in the form of Suggested No Adverse Response Levels (SNARLs). SNARLs are intended as a recommendation of the level of contaminant in drinking water at which adverse health effects would not be anticipated. EPA has adopted RMCLs (MCLGs) for several contaminants that were initially addressed by SNARLs. A new emphasis on issuing MCLGs is intended to eliminate the need for EPA to issue additional SNARLs.

The act also establishes regulations to protect underground drinking water sources by controlling subsurface injection. Five classes of wells are defined, and the requirements which have been promulgated for each class of well are set forth in the regulations. The classes and their associated general requirements are shown in Table 5.7.

Table 5.7
SDWA Well Classifications

Class	Description	Requirements
I	Injects hazardous waste beneath an underground source of water	Permit required to construct, must meet specified construction requirements
II	Operated in conjunction with oil & gas production, enhanced recovery, or storage	Permit required to construct, must meet specified construction requirements
III	Used in mining or recovery of non-fossil fuel	Permit required to construct, must meet specified construction requirements
IV	Injects hazardous materials above or into a substrate containing an underground source of drinking water	Specific requirements deferred to state programs and RCRA, new construction banned
V	All injection wells not described above	States required to take action if well poses risk to human health

In order to protect ground water from underground injections, which endangers drinking water sources, the SDWA directs EPA to do the following:

1. Publish minimum national requirements for effective state Underground Injection Control (UIC) programs;
2. List states that need UIC programs;
3. Make grants to states for developing and implementing UIC programs;
4. Review proposed state programs and approve or disapprove them; and
5. Promulgate and enforce UIC programs in listed states if the state chooses not to participate or does not develop and operate an approvable program.

However, regulations promulgated under the UIC program are not to interfere with or impede oil and gas production (unless necessary to protect underground sources of drinking water) or disrupt effective existing state programs unnecessarily, and are to take local variations in geology, hydrology, and history into account.

The underground injection control program is intended to be administered by the states. Existing Class IV wells must be phased out of operation within six months after a state UIC program is approved, and no new Class IV wells will be permitted. Class IV wells are also addressed by the Resource Conservation and Recovery Act (RCRA), which calls for their elimination through 1984-enacted amendments. Classes I, II, and III wells must be reevaluated and require new state permits every five years.

The 1986 amendments require EPA to identify monitoring methodology for Class I wells, which will provide the earliest possible detection of fluid migration from injection wells toward underground sources of drinking water. Also, EPA must report to Congress on the numbers, categories, and contamination problems associated with Class V wells.

In response to other mandates contained in the 1986 amendments, EPA has prohibited the use of lead solders or flux with a lead content of more than 0.2 percent in public water systems and connected buildings, including residences. The new regulations require public water supply systems to begin notifying customers in June 1988 of any possible lead contamination and potential health effects.

Other provisions of the SDWA which may be related to hazardous materials management address "sole source aquifers" and imminent hazards. Sole source aquifers are underground water supplies that serve as the only source of drinking water for a specific area. Such aquifers may be subject to more stringent protective measures than those applicable to other groundwater sources.

Section 1431 of the SDWA addresses imminent hazards by authorizing EPA to commence "a civil action for appropriate relief, including a restraining order or permanent or temporary injunction" where it has been determined that a contaminant is entering or is likely to enter a public water system, and that appropriate state and local authorities have not acted.

Enforcement

The act establishes a joint federal-state system for assuring compliance with regulatory standards. States are intended to administer the act subject to approval by EPA. However, facilities on Indian lands remain under federal jurisdiction. Failure to enforce federal standards may result in a loss of EPA public drinking water grants.

Upon federal authorization, states issue permits for UIC facilities. A list of states authorized to issue such permits is shown in Table 5.8.

The 1986 amendments strengthened the enforcement alternatives available to EPA by allowing the agency to enter into enforcement action sooner and by increasing the maximum civil penalty from $5,000 to $25,000 per day. Additionally, anyone found guilty of introducing contaminants into public water sources or otherwise endangering public health can be fined $50,000 and imprisoned for five years.

Table 5.8
UIC State Programs

State	Class I Well	Class II Well	Class III Well	Class IV Well	Class V Well
Alabama	X	X	X	X	X
Arkansas	X	X	X	X	X
Alaska*	-	X	-	-	-
California*	-	X	-	-	-
Colorado*	-	X	-	-	-
Connecticut	X	X	X	X	X
Delaware	X	X	X	X	X
Florida*	X	-	X	X	X
Georgia	X	X	X	X	X
Guam	X	X	X	X	X
Idaho	X	X	X	X	X
Illinois	X	X	X	X	X
Kansas	X	X	X	X	X
Louisiana	X	X	X	X	X
Maine	X	X	X	X	X
Maryland	X	X	X	X	X
Massachusetts	X	X	X	X	X
Mississippi*	X	-	X	X	X
Missouri	X	X	X	X	X
Nebraska	X	X	X	X	X
New Hampshire	X	X	X	X	X
New Jersey	X	X	X	X	X
New Mexico	X	X	X	X	X
North Carolina	X	X	X	X	X
North Dakota	X	X	X	X	X
Ohio	X	X	X	X	X
Oklahoma	X	X	X	X	X
Oregon	X	X	X	X	X
Rhode Island	X	X	X	X	X
South Carolina	X	X	X	X	X
South Dakota*	-	X	-	-	-
Texas	X	X	X	X	X
Utah	X	X	X	X	X
Vermont	X	X	X	X	X
Washington	X	X	X	X	X
West Virginia	X	X	X	X	X
Wisconsin	X	X	X	X	X
Wyoming	X	X	X	X	X

* Partial primacy

Relationship to Other Regulations

Regulation of Class I and IV wells is provided under RCRA, and the heavy metals which are included in primary drinking water standards are the same parameters that are specified for toxicity testing in RCRA. Primary drinking water standards set MCLs for certain types of radioactivity. Water quality standards are used to set the concentration limits for contaminants found in groundwater at hazardous waste landfill facilities regulated by RCRA and are also used to determine the level of cleanup required at Superfund sites.

Potential Future Developments

The 1986 SDWA amendments direct EPA to perform certain regulatory activities over the following three years, and it is expected that numerous regulations pertaining to drinking water will subsequently be proposed and adopted. As a result, the late 1980s will probably see significant advances in water treatment technology in response to increased regulatory requirements. Whether EPA can meet Congress' requirements, and how these laws will affect utilities across the country, remains to be determined.

Additionally, EPA has announced an intention to extend the applicability of the SDWA regulations to a class of water distribution systems smaller than those currently regulated. Thus, future regulations may apply to such systems as wells that supply water to rural schools or factories with their own supplies.

6

TOXIC/HAZARDOUS MATERIALS REGULATIONS

As crude a weapon as the cave man's club, the chemical barrage has been hurled against the fabric of life.

—Rachel Louis Carson, *Silent Spring*, 1962

Ironically, enactment of the air and water quality regulations discussed in the preceding chapter contributed to the need for the regulations discussed in this chapter. First, these laws helped to focus a critical eye on examples of environmental quality degradation, such as massive fish kills, public health effects of contaminated drinking water, and visible industrial atmospheric discharges. Second, they actually increased the amount of waste not under the regulatory umbrella.

When air and water quality regulations came into effect, many pollutants that were routinely being discharged into the air or water were subsequently placed in landfills, or simply dumped in remote areas. Many of the pollutants that were subjected to required treatment technologies generated by-products requiring disposal. For example, the widespread adoption of a water treatment technology such as coagulation resulted in vast amounts of sludge, which required some type of disposal. The easiest, most common disposal method for these substances was generally land disposal. However, the public was becoming more aware of subtle, less visible examples of resulting environmental degradation, such as groundwater pollution. And although reducing pollution from point sources was certainly beneficial, it became increasingly apparent that less easily regulated non-point sources are a significant portion of environmental pollution. There was also the question of product versus pollutant in cases where exposure to non-waste materials had as great or greater a risk as exposure to wastes.

Obviously, comprehensive environmental issues still remained to be addressed. The regulations discussed in this chapter were the response to these concerns.

HAZARDOUS MATERIALS TRANSPORTATION ACT (HMTA)

The Department of Transportation (DOT) regulations establish a labeling system applicable to hazardous materials, require the display of placards on land carriers transporting hazardous cargo, and mandate the use of a uniform manifest in the shipment of hazardous waste. The laws also identify stringent specifications which describe the types of packages and containers that can be used to ship hazardous materials. If shipments are to be made by air, passenger rail, or water, additional restrictions apply.

Background

The first regulations pertaining to the transport of hazardous materials were drafted soon after the Civil War to control the transport of explosives by train. Since that time, there has been an increasing need to regulate transporters of hazardous materials to ensure safe and proper delivery to their authorized destination.

Estimates for the amount of hazardous materials transported over U.S. highways range from one to four billion tons per year. It is further estimated that about one in ten trucks on the roadways carries hazardous materials such as petroleum products, chemicals, radioactive materials, and hazardous wastes. About 40 percent of the hazardous materials transported in the United States are estimated to be carried by air, water, pipeline, and rail, as opposed to motor vehicles.

Current transportation laws are based on the HMTA, which was passed in 1975 and is administered by the DOT. EPA promulgated standards for hazardous waste transporters on May 19, 1980, and coordinated these standards with HMTA. DOT views hazardous wastes as a subset of hazardous materials; thus, a hazardous waste is always regarded as a hazardous material subject to DOT regulation. The Resource Conservation and Recovery Act (RCRA) states that EPA's hazardous waste regulations must be consistent with the DOT regulations, which are found in 49 *Code of Federal Regulations* (CFR) 172, 173, 178, and 179.

Content

The hazardous materials table published in 49 CFR 172.101 is frequently referred to as the "heart" of the HMTA regulations. About 16,000 materials and substances are listed, followed by twelve columns that give transport, packaging, and identification requirements.

In this table, DOT identifies stringent specifications describing the types of packages and containers that must be used to ship hazardous materials. The specifications vary depending on the type of material, quantity, and method of transport. Two of the most commonly used containers, steel drums and steel-banded wooden shipping containers, are depicted in Figures 6.1 and 6.2.

Figure 6.1
DOT Specified 17H Steel Drum (30 Gallon)

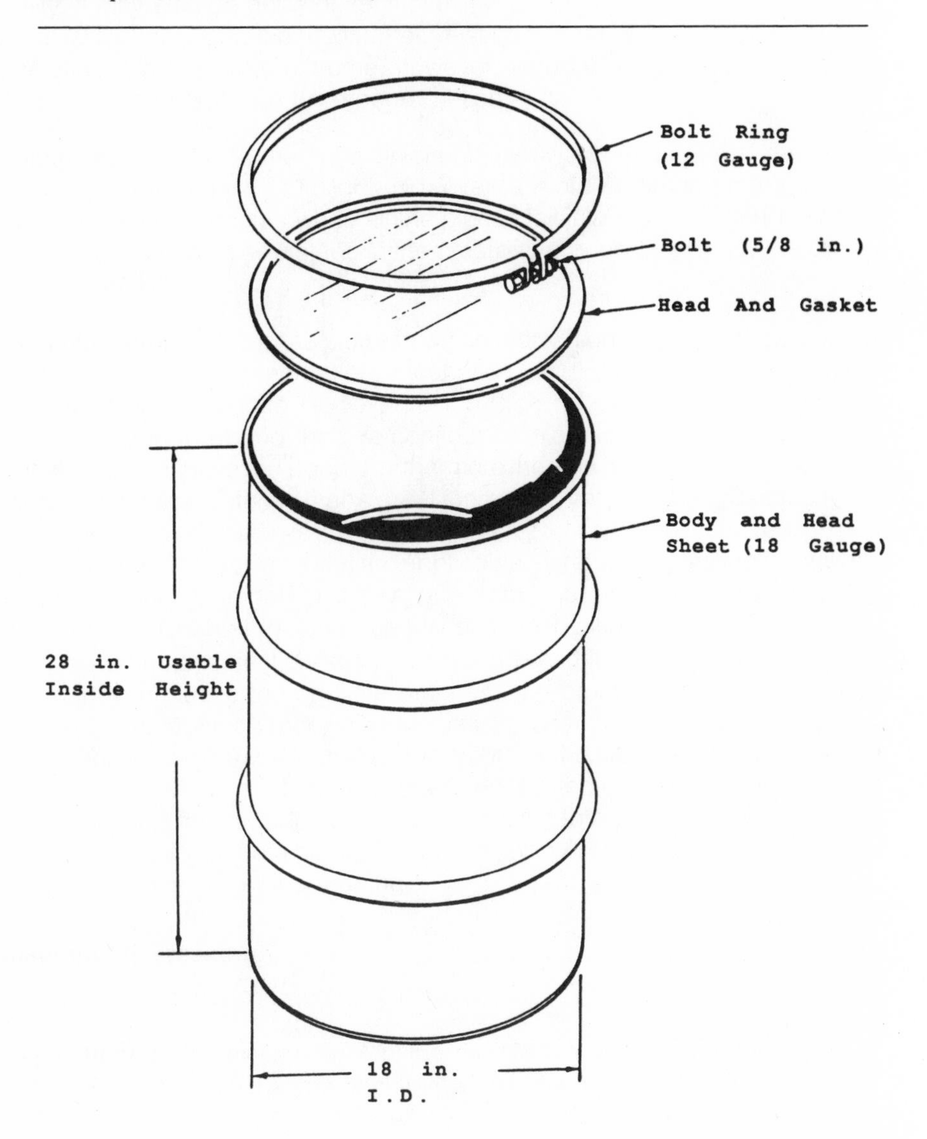

Figure 6.2
DOT Specified 7A Steel-Banded Wooden Shipping Container

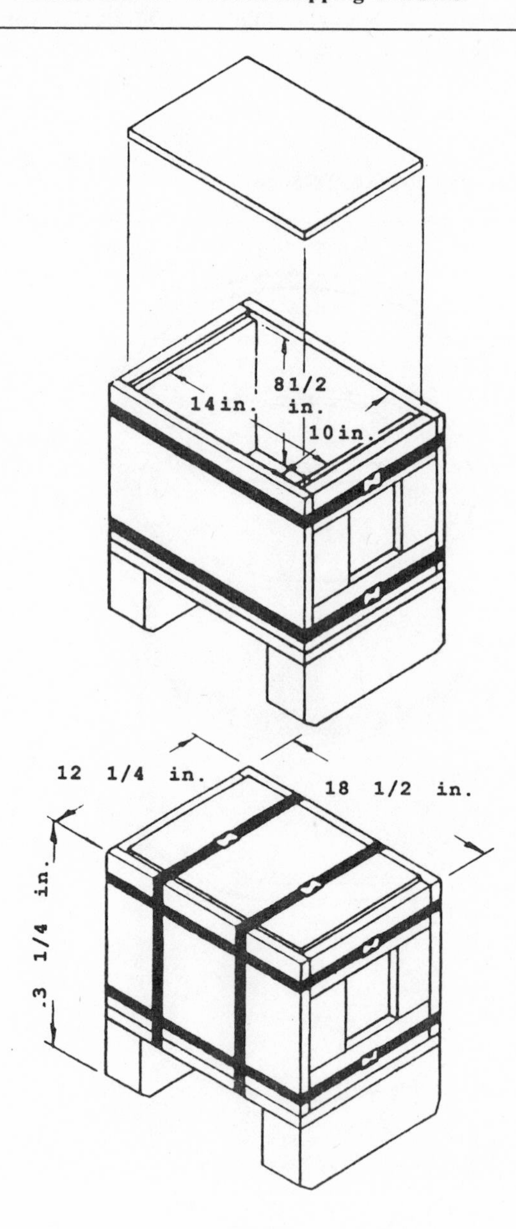

Identifying a hazardous material in accordance with DOT regulations involves marking, labeling, and placarding. Use of all three assures redundancy, and much of the information so conveyed is also contained in the shipping papers that are required for the transport of hazardous materials. The redundancy is to make sure that in case of emergency, vital information is accessible to the driver and is visible on the vehicle and the containers themselves.

Markings are written information which identify the package contents. The most important marking is the shipping name, which is determined by selecting an accurate technical shipping name from the hazardous materials table. If there is not an applicable shipping name, identification must be used that corresponds to the specific hazard class of the material being shipped. In selecting appropriate markings by this system, the highest priority hazard ranking must be used to describe a material if more than one classification is applicable.

A complete list of DOT hazard classifications, definition of each classification, and examples are shown in Table A.5 in the appendix. Table 6.1 lists some common classifications in descending order of priority. This priority system does not include explosives, etiologic agents, or organic peroxides. It does include categories of Other Regulated Materials (ORM). DOT created

Table 6.1
DOT Hazard Classes

Descending Priority:

- Radioactive material
- Poison A
- Flammable gas
- Nonflammable gas
- Flammable liquid
- Oxidizer
- Flammable solid
- Corrosive material (liquid)
- Poison B
- Corrosive material (solid)
- Radioactive material (limited quantity)
- Irritating material (solid)
- Combustible material (in containers exceeding 110 gallon capacity)
- ORM-B
- ORM-A
- Combustible liquid (in containers less than 110 gallon capacity)
- ORM-E

these hazard classes to accommodate those hazardous wastes that are regulated by EPA but do not otherwise fall into one of the previously developed hazard classes. The following is a description of the ORM classes:

ORM-A: A material that has an anesthetic, irritating, noxious, toxic, or other similar property and that can cause extreme annoyance or discomfort to passengers and crew in the event of leakage during transportation.

ORM-B: A material capable of causing significant damage to transporting vehicles if it leaks.

ORM-C: A material that has other inherent characteristics not described as an ORM-A or B, but which makes it unsuitable for shipment, unless properly identified and prepared for transportation. Each ORM-C material is specifically named in the hazardous material table. An example is sawdust.

ORM-D: A consumer commodity that, though otherwise subject to regulation, presents a limited hazard during transportation due to form, quantity, or hazard.

ORM-E: A material not included in any other hazard class but subject to the requirements of the DOT regulations.

EPA's hazardous waste characteristics, as defined by RCRA, do not directly correspond to DOT hazard classes. DOT classifications which are potentially applicable to the Resource Conservation and Recovery Act (RCRA) hazardous waste characteristics are shown in Table 6.2.

Other required markings include those indicating which end of the container is the top and the reportable quantity (RQ), if applicable. If a

Table 6.2
Comparison of EPA's Hazardous Waste Characteristics with DOT's Hazard Classes

EPA Characteristic	DOT Hazard Class
Ignitability	Flammable liquid Combustible liquid Flammable solid Flammable gas Oxidizer
Corrosivity	Corrosive metal
Reactivity	Explosive Flammable solid Organic peroxide Poison B
EP Toxicity	Poison B ORM-A ORM-B

hazardous waste container has a capacity of 110 gallons or less, it must additionally be marked with the following information:

1. Identification number: The identification number is listed in the hazardous materials table and is either a United Nations (UN) or North American (NA) number. These numbers are keyed to emergency response information contained in the *DOT Emergency Response Guidebook*.
2. Name and address of consignee and consignor (except in certain situations)
3. Warning
4. Generator's name and address
5. Manifest document number

No abbreviations may be used except for ORM. For containers greater than 110 gallons, requirements vary and the regulations should be consulted for specific situations.

A label is essentially a sticker of specified design which goes on the package and pictorially indicates the hazard class of the contents. DOT requires labels according to the classification of the hazardous material, and if more than one type of hazard is posed, multiple labels must be used. (A pictorial description of DOT-required labels is provided in 49 CFR 172.) A descriptive summary of the labeling system is presented in Table 6.3. If a

Table 6.3
Descriptive Summary of DOT Labels

Substance	Label
Explosives	Orange with black lettering
Flammable gas	Red with black lettering
Flammable liquid	Red with black lettering
Nonflammable gas	Green with black lettering
Flammable solid	Red and white vertically striped
Corrosive	Diamond shaped, white top, black bottom
Oxidizing substance	Yellow with black lettering
Poisons	Black on white skull and cross bone
Radioactive	Magenta propeller on yellow

package qualifies for DOT's small quantity exclusion listed in the regulations, it does not have to be labeled.

A placard is similar to a label in that it pictorially illustrates the hazard class of the material being shipped. However, a placard is generally displayed on the transport vehicle, while a label is attached to the hazardous material container.

DOT requires the display of placards on land carriers which transport hazardous cargo. The placards are intended to serve as a notice of the potential danger of the cargo before labels on packages are observed or before any shipping papers are read. Placards are intended to convey the same basic identification information as DOT-required labels and are also shown in 49 CFR 172.

The generator of a hazardous waste is required to provide the transporter with proper placards identifying the material being shipped. (Generators are termed "shippers" and transporters are termed "carriers" under DOT regulations.) DOT regulations specify that any quantity of waste having a hazard class rating must be placarded.

Shipping paper requirements for hazardous materials are specified by DOT regulations. Informational content must include proper shipping name, quantities of material, hazard class, and UN or NA identification number. In addition, hazardous waste shipments must be accompanied by a uniform hazardous waste manifest, which is shown in Figure 6.3. The uniform hazardous waste manifest is an EPA document, and DOT authorizes its use as a shipping paper.

Similar to other environmental regulations previously discussed, the HMTA contains an imminent hazard clause, found in section 111 of the act.

Enforcement

Four agencies within DOT have primary regulatory responsibility for enforcement of the regulations:

1. Federal Aviation Administration (FAA)—air transportation
2. Federal Highway Administration (FHWA)—highway transportation
3. Federal Railway Administration (FRA)—railway transportation
4. United States Coast Guard (USCG)—transportation by vessel

In addition, the Materials Transportation Bureau (MTB), within the Research and Special Programs Administration (RSPA), issues and interprets the regulations. The MTB consists of the following two offices:

1. Office of Hazardous Materials Regulations (OHMR)—develops the regulations
2. Office of Operations and Enforcement (OOE)—furnishes guidelines for enforcement, containers manufacturing, drum reconditioning, and radioactive materials

Figure 6.3
Uniform Hazardous Waste Manifest

Please print or type. (Form designed for use on elite (12-pitch) typewriter.)

Form Approved. OMB No. 2000-0404. Expires 7-31-86

UNIFORM HAZARDOUS WASTE MANIFEST

1. Generator's US EPA ID No. Manifest Document No.

2. Page 1 of

Information in the shaded areas is not required by Federal law.

3. Generator's Name and Mailing Address

A. State Manifest Document Number

B. State Generator's ID

4. Generator's Phone ( )

5. Transporter 1 Company Name

6. US EPA ID Number

C. State Transporter's ID

D. Transporter's Phone

7. Transporter 2 Company Name

8. US EPA ID Number

E. State Transporter's ID

F. Transporter's Phone

9. Designated Facility Name and Site Address

10. US EPA ID Number

G. State Facility's ID

H. Facility's Phone

11. US DOT Description (Including Proper Shipping Name, Hazard Class, and ID Number)	12. Containers No.	12. Containers Type	13. Total Quantity	14. Unit Wt/Vol	I. Waste No.
a.					
b.					
c.					
d.					

J. Additional Descriptions for Materials Listed Above

K. Handling Codes for Wastes Listed Above

15. Special Handling Instructions and Additional Information

GENERATOR

16. **GENERATOR'S CERTIFICATION:** I hereby declare that the contents of this consignment are fully and accurately described above by proper shipping name and are classified, packed, marked, and labeled, and are in all respects in proper condition for transport by highway according to applicable international and national governmental regulations.

Printed/Typed Name | Signature | Date: Month Day Year

TRANSPORTER

17. Transporter 1 Acknowledgement of Receipt of Materials

Printed/Typed Name | Signature | Date: Month Day Year

18. Transporter 2 Acknowledgement or Receipt of Materials

Printed/Typed Name | Signature | Date: Month Day Year

FACILITY

19. Discrepancy Indication Space

20. Facility Owner or Operator: Certification of receipt of hazardous materials covered by this manifest except as noted in Item 19.

Printed/Typed Name | Signature | Date: Month Day Year

EPA Form 8700-22 (3-84)

Another group of agencies has jurisdiction over some aspects of hazardous materials transportation. The Department of Energy (DOE) is largely concerned with fuels and the Department of Defense (DOD) with materials used for military purposes. The Nuclear Regulatory Commission (NRC) has jurisdiction over high-level radioactive substances in the civil sector, while EPA has certain transportation-related responsibilities for selected chemicals and hazardous nonnuclear waste.

States and municipalities cannot enforce standards inconsistent with or more stringent than the federal DOT regulations. However, states have recently become more involved because of a concern with public safety, resulting in additional licensing, registration, and permit requirements which may vary widely at the state and local level.

Relationship to Other Regulations

Although section 3003 of RCRA directs EPA to regulate transporters of hazardous waste, EPA's regulations incorporate and rely on the requirements already established by DOT under the HMTA.

Transporter requirements are established by RCRA for certain instances involving spills, mixing of wastes, and storage of wastes. A transporter who selects the disposal site may be strictly and jointly liable under various provisions of the Comprehensive Environmental Response, Compensation, and Liability Act (CERCLA). Both CERCLA and the Clean Water Act (CWA) define substances and reportable quantities (RQs) which, if released into the environment by a transporter, must be reported to the National Response Center (NRC).

Potential Future Developments

The provisions of the HMTA and related regulations have remained substantially unchanged since promulgation of final rules. One significant regulatory development was the requirement for hazardous waste transporters to use a uniform manifest. This requirement became effective in 1984.

The hazardous materials table and its corresponding requirements are revised frequently to add or delete substances or to change requirements applicable to a listed substance.

In late 1987, significant regulatory revisions were proposed, intended to improve safety and make the rules easier to understand. Included in the proposed changes is a new classification system combined with a new packaging system to more appropriately match the hazard with the type of container. The proposed changes would also align U.S. requirements with international standards to facilitate the safe handling of hazardous materials shipped in international commerce. It is likely these proposed changes will be adopted sometime in 1988, especially since international shipments under existing DOT rules will not be accepted by international regulatory bodies after December 31, 1989.

TOXIC SUBSTANCES CONTROL ACT (TSCA)

TSCA regulates existing and new chemical substances and, therefore, primarily applies to ongoing chemical manufacturing operations and the products of those operations. TSCA grants EPA broad regulatory authority over most chemical substances. Pesticides, tobacco, foods, food additives, drugs, cosmetics, firearms, and regulated nuclear materials are the main categories of chemicals which are not within the statute's scope.

The goal of the act is to reduce unreasonable environmental or human health risks from dangerous chemicals. The goal is to be achieved by controls on the manufacture, distribution, use, and/or disposal of such dangerous chemicals.

Background

TSCA was drafted in response to repeated incidents involving adverse health, environmental, and economic damages inflicted by widely used substances such as polychlorinated biphenyls (PCBs), kepone, vinyl chloride, polybrominated biphenyls (PBBs), and asbestos.

National concern with widespread contamination by toxic compounds became evident in the 1960s, and by 1972 the Clean Air Act (CAA) and the Clean Water Act (CWA) had been promulgated to address some of these concerns. However, both these acts dealt with substances only after they had entered the environment as waste, and federal authority to regulate the complete cycle of chemicals from production to disposal was severely limited. TSCA was designed to fill this regulatory gap by granting EPA the authority to require testing before human or environmental exposure.

The task of drafting a federal toxics law was begun in 1971 and completed in 1976. TSCA was enacted on January 1, 1977, exactly seven years to the day after the passage of the National Environmental Policy Act (NEPA) had been heralded by some as "the beginning of an environmental decade."

Content

TSCA regulates the safety of raw materials and, therefore, primarily applies to ongoing chemical manufacturing operations and the products of those operations. TSCA applies to any entity that manufactures, processes, or distributes in commerce any chemical substance or mixture. The terms "manufacture" and "process" are defined broadly by the act. "Manufacture" includes importing and production—even production of a chemical as a by-product. "Process" includes incorporation of a chemical into an article by manufacturers of consumer goods or industrial products. Thus, TSCA can potentially apply to companies outside the chemical industry. However, pesticides, tobacco, foods, food additives, drugs, cosmetics, firearms, and regulated nuclear material are not within the statute's scope.

TSCA has two main regulatory goals. First is the acquisition of sufficient data to identify and evaluate potential chemical hazards. The second goal is to reduce these hazards by controlling the production, use, distribution, and disposal of such substances as deemed necessary. These goals are to be achieved through four major activities:

1. Screening of new chemicals;
2. Testing of chemicals identified as potential hazards;
3. Gathering of information on existing chemicals, including the development of systems to evaluate the data and the specification of recordkeeping and reporting requirements; and
4. Controlling chemicals proven to pose a hazard.

Virtually all new chemicals must undergo screening and review prior to manufacture. In May 1979, EPA published its initial inventory of chemical substances. Any chemicals not listed were considered new chemicals subject to premanufacture review as of July 1979. As new chemicals complete premanufacture review and begin to be manufactured they are added to the inventory.

If a chemical is not already listed on the inventory, a premanufacturing notice (PMN) must be submitted. This notification must, among other things, identify the chemical, provide information on use, method of dispersal, production levels, worker exposure, and potential by-products or impurities. In addition, the manufacturer must provide any data possessed on health and environmental effects of the product and a description of known or reasonable ascertainable data. After the submittal of the PMN, EPA has 90 days to complete the review and either approve the production of the chemical or act to ban manufacture.

If a chemical is approved for manufacture, EPA may propose a Significant New Use Rule (SNUR). Under this rule the use or production volume of a chemical may be restricted. Because this provision has been opposed by the chemical industry, and because determination of a new use can be difficult, the SNUR provisions have not been widely applied. Historically, SNURs have been issued on an ad hoc basis.

Under section 4 of the act, testing can be required for a substance with insufficient test data to evaluate its effects if either of the following conditions exist: (1) The substance could pose an unreasonable risk to health or the environment; or (2) The substance is produced in large volumes and could enter the environment in substantial quantities or be in contact with humans in significant amounts.

Chemicals are selected for testing by an interagency testing committee composed of representatives from eight federal agencies. Once selected, these chemicals or groups of chemicals are placed on a priority list for testing which may never contain more than 50 chemicals or chemical groups.

Because toxicity testing is so expensive (up to $1,000,000/chemical), TSCA requires that with minor exceptions all manufacturers and processors of the material must share the costs of testing. The allocation of costs is to be devised by the manufacturers or if they cannot agree, by EPA.

Section 8 of TSCA gives EPA the authority to require chemical manufacturers and processors to gather information and submit certain data. Failure to submit the requested information can be judged a civil or criminal offense, with related fines and jail sentences.

EPA developed the Chemicals-in-Commerce Information System (CICIS) to store and retrieve TSCA data. The computerized TSCA inventory became operational in late 1979, and several information services have derived from it, including subsystems for Freedom of Information Act requests, inventory profiles for EPA regional offices, support for the TSCA premanufacture review process, and health and safety study submissions.

The Interagency Toxic Substances Data Committee (ITSDC), formed in February 1978 by EPA and the Council on Environmental Quality (CEQ), is continuing its work to construct a comprehensive Chemical Substances Information Network (CSIN). CSIN enables users of toxic substances information to have access to a number of independent and autonomous data banks in the public and private sectors. The goal of CSIN, which became operational in November of 1981, is to enhance the availability of chemical data to both governmental and private-sector organizations to efficiently resolve and manage issues concerning chemical substances.

TSCA defines stringent recordkeeping and reporting requirements to generate data on current product usage effects. Standard TSCA requirements include:

1. Reporting of production and exposure data on specified chemicals;
2. Retention of reports of significant adverse reactions;
3. Submission of health and safety studies related to certain listed chemicals;
4. Notice of substantial risk by chemical manufacturers, processors, or users who become aware of a substantial threat from a chemical; and
5. Records concerning health effects on employees must be kept for 30 years. Other records must be retained for five years.

In November 1984, EPA set forth general exemption standards for manufacturers of small quantities of chemical substances. Manufacturers that qualify under these standards may be exempt from some of the reporting and recordkeeping rules.

Control measures granted to EPA include:

1. Limiting allowed uses of chemical;
2. Requiring special labeling;
3. Requiring special use instructions;
4. Requiring special recordkeeping;
5. Setting disposal standards; and
6. Requiring quality control procedures in chemical manufacturing processes.

Section 6 of TSCA provides extensive authority for the control or banning of specific substances that are found to pose an unreasonable risk to human health or the environment. Allowable regulatory actions range from labeling requirements to complete prohibition. PCBs, chlorofluorocarbons (CFCs), asbestos in schools, and dioxins have been regulated under this provision.

PCBs are of concern because tests on laboratory animals show that chronic exposure to PCBs may cause reproductive failures, gastric disorders, skin lesions, and tumors. Moreover, PCBs persist and, when released into the environment, tend to accumulate in tissues of living organisms. As PCBs move up in the food chain, their concentration, and thus the potential for detrimental health effects, increases. TSCA has essentially prohibited the use of PCBs since January 1, 1978. Exceptions are for applications that are either "totally enclosed and pose an insignificant risk of human exposure," or for applications that EPA determines "will not present an unreasonable risk of injury to health and the environment."

CFCs are widely used as refrigerants, coolants, industrial solvents, and as a component in some foam packaging. Concern regarding their use is based on studies which conclude that they migrate upward to destroy the stratospheric ozone layer that helps shield humans from the sun's ultraviolet rays. TSCA deals with CFCs by basically outlawing most aerosol uses.

Asbestos is mineral material, which has been widely used in the building industry, and has been demonstrated to be carcinogenic. TSCA requires inspection for exposure of students to airborne asbestos fibers. Asbestos regulation is discussed further in Chapter 7.

Dioxin is an extremely carcinogenic chemical that has been identified as a component in some herbicide production wastes. Waste dioxins were initially regulated under section 6 of TSCA, but were generally turned over to regulation under the Resource Conservation and Recovery Act (RCRA) in 1985. However, proposed regulations were promulgated the same year under section 4 of TSCA, requiring specified manufacturers to test their products for dioxin contamination.

Similar to other environmental laws, TSCA contains an imminent hazard provision. EPA is authorized to take "appropriate action in relation to a chemical substance or mixture which presents an imminent and unreasonable risk of serious or widespread injury to health or the environment." These actions can include seizing the product or seeking judicial action to have production and/or distribution of the product frozen.

Enforcement

TSCA is enforced by EPA through the Office of Pesticides and Toxic Substances (OPTS), and authorized enforcement actions include inspection, penalties, and seizure of substances manufactured in violation of the act.

Unlike many other federal environmental regulations, TSCA has no provision that authorizes state agencies to undertake program enforcement. However, EPA is actively committed to a policy of delegating authority and decision-making power to the states to implement TSCA.

With certain exceptions, TSCA will not affect the authority of any state or political subdivision to establish regulations concerning chemicals. If EPA issues a testing requirement for a chemical, a state may not establish a similar one for the same purposes. However, if EPA restricts the manufacture or otherwise regulates a chemical under TSCA, a state may only issue requirements that are identical, are mandated by other federal laws, or prohibit the use of the chemical.

In response to a request by a state, EPA may grant an exemption to allow the state to regulate differently from TSCA regulatory actions under certain conditions. Specifically, EPA can grant exemptions if the state requirement would not cause a person or activity to be in violation of a requirement under TSCA, and would provide a greater degree of protection and not unduly burden interstate commerce.

Additionally, TSCA authorizes grants to be awarded to states to help establish programs to prevent or eliminate unreasonable risks associated with toxic substances.

Relationship to Other Regulations

EPA is not authorized to proceed under TSCA to control a risk that can be adequately dealt with by actions taken under another statute. Any other federal statute having purview of a chemical substance may take precedence over TSCA. For example, section 112 of the CAA and section 307 of the CWA also specifically address toxic substances. The provisions of these acts are to take precedence over provisions contained in TSCA, unless EPA determines it is in the public interest to invoke TSCA.

Similarly, EPA's exercise of authority under TSCA does not constitute a preemption of jurisdiction under the Occupational Safety and Health (OSH) Act. The Occupational Safety and Health Administration (OSHA) has responsibility for worker protection standards under National Institute for Occupational Safety and Health (NIOSH) standards.

The Superfund Amendments and Reauthorization Act (SARA) requires the development of certain research programs, some of which will be accomplished using TSCA authority. Certain test information required under the Resource Conservation and Recovery Act (RCRA) to identify wastes that may pose a substantial hazard to human health and the environment will be obtained in the same manner.

Substances not regulated by TSCA may be regulated by the Federal Food, Drug, and Cosmetics Act (FFDCA); or the Federal Insecticide, Fungicide, and Rodenticide Act (FIFRA).

Potential Future Developments

It appears unlikely that TSCA-related legislation will be passed in the near future because of its perceived low priority in relation to other environmental bills. However, TSCA grants widespread potential regulatory powers to EPA, and the importance of TSCA is likely to expand in the long run. Particularly as public concern over possible chemical risks and hazards increases, legislators may become more aware of the enormous information-generating and data-collection powers that can be invoked through TSCA.

In a recent regulatory development, EPA proposed in 1987 to issue a generic SNUR which would apply to all substances in certain regulatory categories. The intent of this rule is to establish an expedited process for follow-up on certain new chemical substances. Other areas of active regulatory development include a policy for nationwide PCB cleanup standards and potential bans on nonaerosol use and production of CFCs.

RESOURCE CONSERVATION AND RECOVERY ACT (RCRA)

RCRA regulates the handling of hazardous waste at currently operating or future facilities and is intended to provide for the environmentally sound disposal of waste materials. Thus, it addresses a gap left by the Clean Air Act (CAA) and the Clean Water Act (CWA), which only require that industry remove hazardous substances from air emissions and water discharges.

RCRA establishes the five major elements characterizing the federal approach to hazardous waste management:

1. Classification of hazardous waste
2. Cradle-to-grave manifest system
3. Standards for generators, transporters, and facilities which treat, store, or dispose of hazardous waste
4. Enforcement through a permitting program
5. Authorization of state programs to operate in lieu of the federal program

EPA staffs a toll-free line to answer RCRA-related questions. The number is 1-800-424-9346.

Background

RCRA traces its beginnings to the Solid Waste Disposal Act (SWDA), which was passed in 1965. In 1970 the act was amended to require an investigation of hazardous waste management activities in the United States. This concern with hazardous waste disposal practices was fueled by passage of the CAA and CWA, since sludge and other debris generated by air and water pollution control equipment required disposal guidelines.

Statistics pertaining to the generation of hazardous waste have demonstrated that in sheer size alone the management challenge is awesome. Estimates of the rate of hazardous waste generation range from one pound per person per day to 298 million tons per year in the United States. Additionally, EPA has concluded that as much as 90 percent of hazardous waste has historically been managed by practices which were not adequate for protection of human health and the environment.

RCRA was passed on October 21, 1976, replacing the SWDA. Development of the congressionally mandated regulatory program was subject to various delays. The first hazardous waste rules were issued in February 1980, beginning what has become one of the largest and most controversial regulatory programs.

At the beginning of 1984, hazardous waste regulations occupied over 250 pages of the *Code of Federal Regulations* (CFR), and the number and length of RCRA regulations has substantially increased in every subsequent year. Additionally, significant amendments extending the scope of the original act were promulgated in November 1984. These amendments are the major driving force behind current regulatory developments.

Content

RCRA has three subtitles which create significant regulatory programs. Subtitle D addresses the disposal of nonhazardous solid waste. Subtitle C authorizes a comprehensive federal program to regulate hazardous wastes from generation to ultimate disposal, known as cradle-to-grave regulation. This discussion will deal mainly with Subtitle C, which establishes procedures for waste classification, a cradle-to-grave manifest system, regulatory standards, and permit requirements. Subtitle I resulted from 1984 amendments to the original act and addresses underground storage tanks containing regulated substances other than wastes.

One of the major elements of RCRA is classification of hazardous waste. Part 261 of RCRA defines the criteria for determining whether a waste is hazardous and, therefore, subject to regulation. In order to be regulated, a waste must be both a solid waste and a hazardous waste. The actual physical state of the waste means little according to the act: certain liquids, sludges, or contained gases are solid wastes by RCRA definitions, as shown in Figure 6.4. Substances generally deemed not to be solid wastes include domestic sewage, CWA regulated point discharges, irrigation return flows, radioactive material which is regulated by other agencies, and in-situ mining wastes.

Figure 6.5 presents a general schematic decision tree for determining if a RCRA solid waste is hazardous. A solid waste is usually subject to hazardous waste regulations if it exhibits any of the four RCRA-defined character-

Figure 6.4
Identification of a RCRA Solid Waste

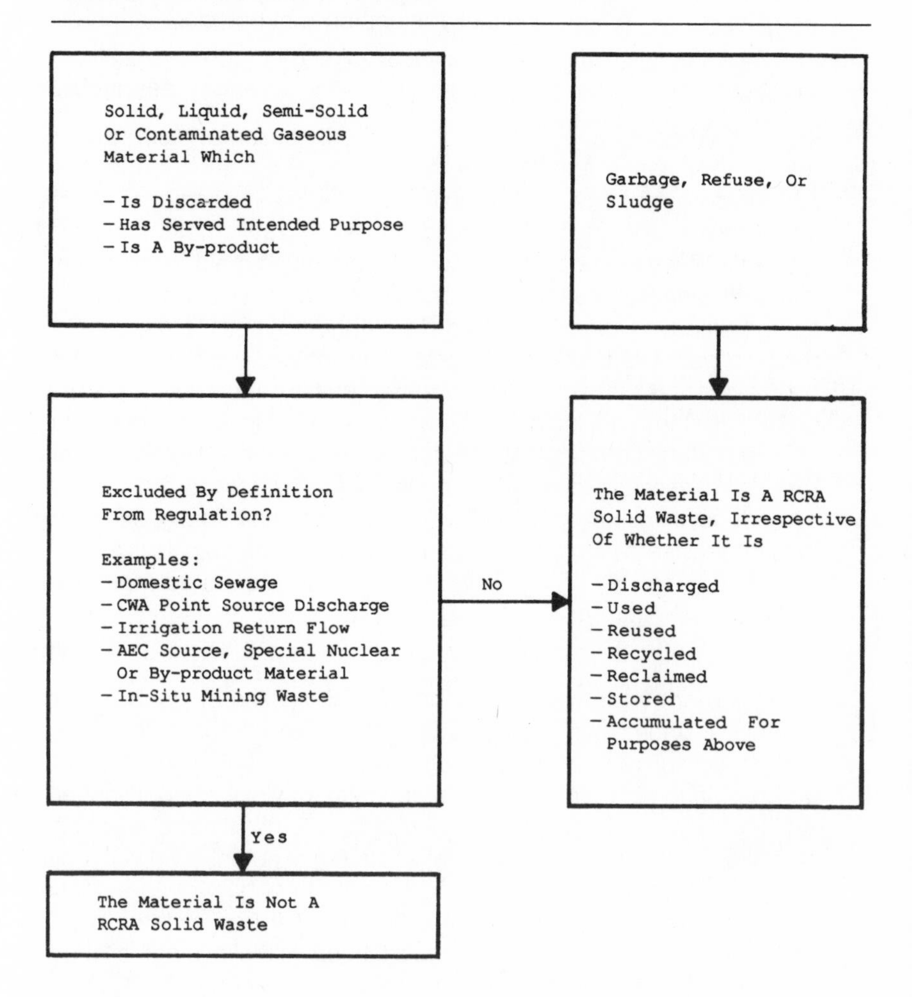

istics of a hazardous waste, if it is specifically listed in the act as being a hazardous waste, or is a mixture of a listed hazardous and nonhazardous waste. RCRA hazardous waste characteristics are as follows:

Ignitable—flash point greater than 1400° F.

Corrosive—pH less than or equal to 2 or greater than or equal to 12.5.

Extraction Procedure (EP) Toxicity—100 × selected drinking water standards.

Reactive—gives off fumes, is explosive, or reacts violently with water.

Figure 6.5
Identification of a RCRA Hazardous Waste

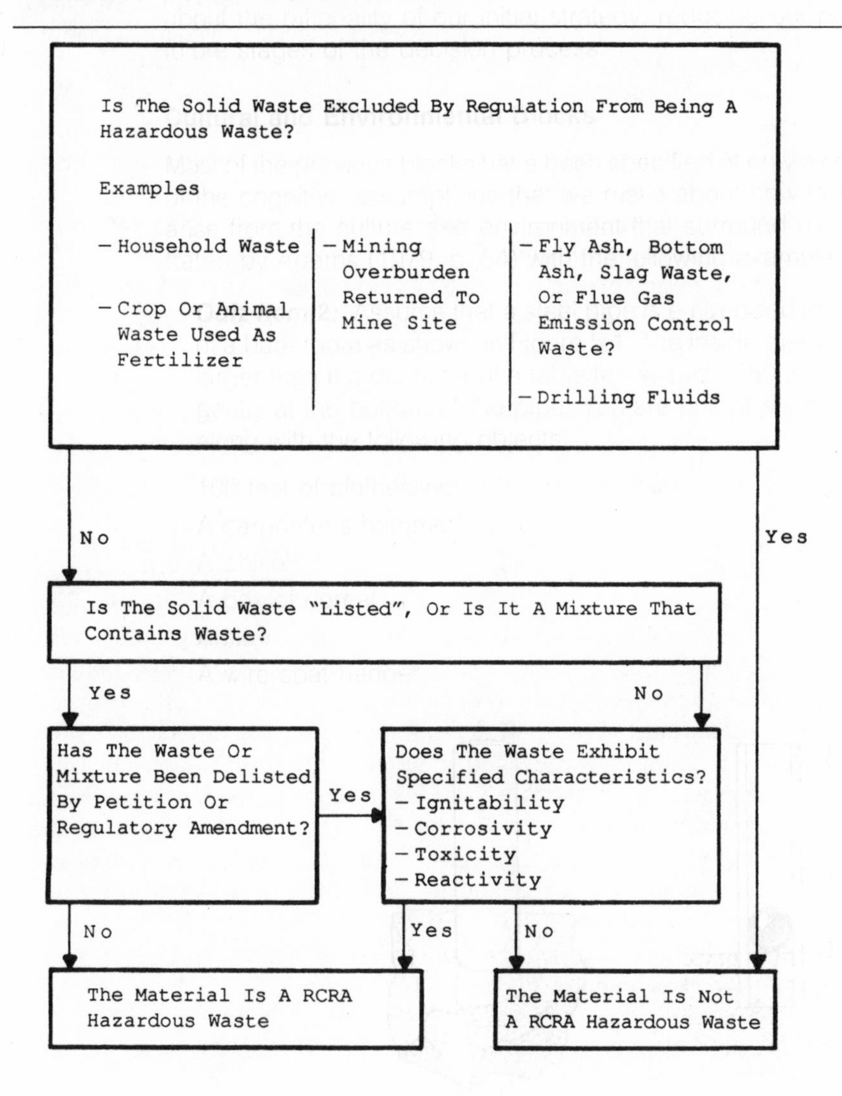

A waste may be listed in the act by name or source. Table A.6 in the appendix lists hazardous wastes from specific sources and the corresponding EPA hazardous waste number. Some listed wastes are defined as acutely hazardous, and special, more stringent rules apply to these wastes. Table A.7 lists RCRA acute hazardous wastes and corresponding EPA hazardous waste numbers. Other categories of listed wastes include hazardous wastes from nonspecific sources and toxic wastes, shown in Tables A.8 and A.9, respectively.

Exemptions include industrial waste-water discharges; nuclear materials; fly ash; mining overburden; drilling fluids; wastes generated by small quantity generators (SQGs); and wastes generated by activities associated with recycling, reuse, or reclamation of waste materials. Materials that are reused or recycled may or may not be regulated as a hazardous waste. The determination is essentially made case-by-case and may depend on the reuse function, whether the waste is a listed or characteristic hazardous waste, and the specific chemical or trade identification of the substance being reused or recycled. Also, a material defined by the regulations as a hazardous waste can be exempted from regulation through delisting by petition or through regulatory amendment.

The cradle-to-grave manifest system enables EPA to track hazardous waste from the point of generation to the point of final disposal, utilizing standards which are applicable to generators and transporters of hazardous waste. These requirements are summarized in Table 6.4

Generators must have an EPA identification number and must ensure that wastes are shipped in proper containers, are accurately labeled, and are accompanied with proper placards for use by the transporter. Generators are generally required to ship their wastes offsite within 90 days after beginning accumulation, or they must have a storage permit and comply with applicable storage standards for containers and tanks. Exceptions, both more and less stringent, may be applicable to SQG satellite facilities and acutely hazardous wastes. Generators must also conduct proper operating, maintenance, and inspection procedures; conduct personnel training; and prepare a contingency plan to be followed in the event of an emergency.

Transporters must comply with DOT regulations according to HMTA and with the hazardous waste manifest system requirements shown in Table 6.4. If the generator has not received a return copy of the hazardous waste transport manifest within 35 days of the date the waste was accepted by the initial transporter, the generator must contact the transporter and/or the designated Treatment, Storage, and/or Disposal Facility (TSDF) to determine the status of the waste. If the generator still does not receive a return copy of the manifest, an exception report must be filed within 45 days of the date the waste was accepted by the initial transporter.

All TSDFs must eventually comply with specified standards. Requirements are dependent on the type of facility and whether the facility is cur-

Table 6.4
Generator and Transporter Standards

Generator Requirements:

- -Identify waste(s)
- -Notify EPA
- -Obtain identification number(s)
- -Utilize manifest system
- -Observe proper packing system
- -Ship to treatment, storage, or disposal facility (TSDF)
- -Conduct inspections
- -Conduct personnel training
- -Prepare required reports
- -Store waste less than 90 days (or obtain permit)
- -Prepare contingency plans

Transporter Requirements:

- -File notice with EPA
- -Obtain identification number(s)
- -Use manifest
- -Delivery to TSDF
- -Maintain and retain proper records
- -Discharges: take appropriate action and report to DOT

rently under operation or is defined as a new facility. Facilities which were in operation or under construction on November 19, 1980, may operate under interim status standards until a final permit is obtained from EPA.

Standards for TSDFs can be categorized as general or specific. General requirements applicable to virtually all TSDFs are shown in Table 6.5. The owner or operator of the facility must also analyze wastes, provide security, conduct inspections, provide employee training, maintain emergency equipment, and formulate an emergency response plan.

TSDFs must have a written plan to provide for closing the site at the end of the life of the facility. Proof of financial resources to be used for site closure must be shown by a guarantee. This guarantee may be in the form of a trust fund, a surety bond, letter of credit, or insurance policy. Other types of financial guarantees may be acceptable in certain circumstances.

Generators and TSDFs must keep all required records for three years according to federal regulations. Training records for current personnel are to be kept until the facility closes.

Specific standards apply to designated categories of TSDFs. Containers; tanks; surface impoundments; waste piles; land treatment units; landfills; incinerators; thermal treatment units; chemical, physical, and biological treatment units; and underground injection wells are all specifically regulated.

Table 6.5
General TSDF Requirements

-Notify EPA
-Obtain interim status (if applicable)
-Observe facility standards
-Institute record-keeping
-Prepare required reports
-Obtain permit
-Formulate closure plan
-Obtain financial assurance for closure

RCRA-mandated facility-specific standards have been extremely controversial and subject to change. Many regulatory changes and additional requirements have occurred in response to the 1984 amendments to RCRA. However, groundwater monitoring is one type of specific standard that is applicable to many TSDFs.

At least one upgradient and three downgradient wells are usually required, and specific parameters to be analyzed vary depending on permit requirements. Generally, samples from each well in the system must be analyzed quarterly for all drinking water standards, groundwater quality parameters, and indicator parameters during the first year the facility operates. After the first year, groundwater parameters must be analyzed annually. Indicator parameters must be analyzed semi-annually. Once compliance monitoring is initiated as part of a permit requirement, it must be continued for the active life of the facility. Less stringent monitoring may be allowed during post-closure care if favorable conditions are demonstrated.

All TSDFs are required to obtain a RCRA permit, which consists of two parts: part A of the application requires a minimal amount of general information, while Part B is a complex engineering report, requiring extensive site-specific information and substantial documentation. Permit conditions are mainly determined on a case-by-case basis. Permit amendments, modification procedures, and rules applicable to expired permits are addressed in the act.

RCRA was reauthorized in 1984 by a law officially titled the Hazardous and Solid Waste Amendments of 1984 (HSWA). These amendments added 72 specific provisions to RCRA and directed EPA to enact 58 of them by 1986. Under the HSWA, RCRA became the primary vehicle for regulating underground storage tanks (USTs). Because of the possible adverse impacts on ground and surface water, leaking underground storage tanks (LUSTs) have emerged as one of the leading environmental concerns of the 1980s. Standards for USTs containing wastes were already addressed in RCRA's subtitle C as facility-specific standards. The HSWA extended RCRA's regulatory authority to include regulated substances other than wastes. These standards are found in subtitle I of RCRA.

An underground storage tank is defined to include all tanks containing regulated substances where the tank volume, including piping, is 10 percent or more beneath the surface of the ground. Thus, numerous tanks which appear to be above ground come under regulation, particularly if there is extensive underground piping. Regulated substances include any substance defined in section 101(14) of the Comprehensive Environmental Response, Compensation, and Liability Act (CERCLA), in addition to petroleum, including crude oil or any fraction thereof.

As with the determination of hazardous wastes, numerous exclusions exist. These include, but are not limited to, farm or residential tanks less than a threshold capacity and used for storing motor fuel for noncommercial purposes, tanks used for storing heating oil on the premises where used, septic tanks, pipelines, surface impoundments, and storm/wastewater collection systems.

The Subtitle I regulatory program consists of notification requirements and technical standard requirements. All UST owners are required to notify EPA or the authorized state agency regarding tank age, size, type, location, and use. Technical standards for the installation and operation of USTs were proposed in 1987. Under these standards, new petroleum tanks will be required to have a protected single wall with a leak detection system and an overfill and spill collection system. In addition, all new chemical tanks will be required to have secondary containment. Existing tanks must be tested for leaks and replaced or upgraded to meet the new standards within ten years.

Two other significant provisions set forth in the HSWA included banning land disposal for specific hazardous wastes and reducing the SQG exemption classification by an order of magnitude. Additional concerns addressed by the HSWA included technical standards for various TSDFs, burning and blending of wastes, interim status facilities, and citizen suits.

Similar to other environmental statutes, RCRA contains an imminent hazard provision. It states that EPA may take action against persons whose actions, past or present, create an imminent or substantial endangerment to health or to the environment.

Enforcement

RCRA is intended to be administered at the state level, upon final authorization by EPA, and permits are issued by state regulatory agencies where state programs have been authorized to act in lieu of the federal program. To receive final authorization, a state program must be "equivalent to and consistent with" the federal program and must provide for adequate enforcement. States may regulate more materials than are regulated by EPA and may impose more stringent environmental standards. As of October 1986, 40 states, the District of Columbia, and Guam had final authorization for the pre-HSWA RCRA program.

RCRA enforcement provisions, which are numerous and substantial, include the definition of "knowing endangerment," which is a felony. This offense can result in fines up to $250,000 for individuals and up to $1 million for corporations, in addition to imprisonment for up to five years. Other RCRA violators are subject to civil penalties of up to $25,000 per day, and criminal offenders are subject to penalties of $50,000 per day and two-year imprisonment.

Subtitle D of RCRA is at the opposite end of the regulatory spectrum from TSCA, which provides only for federal enforcement. Other than publishing criteria for sanitary landfills and inventory of open dumps, EPA has little to do with the regulations of nonhazardous solid waste disposal. Generally, nonhazardous waste disposal programs are conducted by state and local agencies without federal involvement.

Relationship to Other Regulations

The 1986 amendments to Superfund caused RCRA to be amended. CERCLA partially defines the substances which are regulated under Subtitle I of RCRA if contained in USTs. The treatment, disposal, and transportation of hazardous wastes involved in the cleanup of a Superfund site generally requires compliance with RCRA-specified standards, but does not require a RCRA permit.

RCRA-defined hazardous wastes are considered a subset of hazardous materials and, as such, are subject to DOT requirements for packaging, marking, labeling, placarding, spill reporting, and shipping paper preparation.

RCRA currently regulates Class I and IV type underground injection wells as defined by the Safe Drinking Water Act (SDWA). The RCRA-defined extraction procedure (EP) toxicity test requires analysis for parameters identified by the SDWA as primary drinking water standards.

Potential Future Developments

Due to extensive amendments passed in November 1984, significant additions to RCRA are being formulated, and previously promulgated rules are being revised. The content of many of the final rules remains to be determined and ultimately may be resolved through judicial interpretation in numerous cases. Major issues likely to be addressed in the near future include the feasibility of significant bans on land disposal, clarification of terminology describing exemptions to UST regulations, and finalization of technical standards for USTs.

A new direction in regulatory policy may be indicated by rules proposed in 1987, requiring new hazardous waste land disposal facilities to have a leak-detection system. This represents a significant move away from groundwater monitoring as the primary means of detecting leaking hazardous waste. The rules are intended to prevent the leakage of hazardous waste into groundwater, eliminating the requirements for costly cleanup measures.

COMPREHENSIVE ENVIRONMENTAL RESPONSE, COMPENSATION, AND LIABILITY ACT (CERCLA)

CERCLA grants the federal government the authority to undertake activities under the directives of a National Contingency Plan (NCP) in order to clean up dangerous, inactive disposal sites and emergency spill situations. Included is the authority to carry out investigations, testing, and monitoring of disposal sites, in addition to implementing remedial measures. Financial support for these actions is provided by the Hazardous Substance Response Trust Fund (commonly termed the "Superfund"). Financial responsibility for cleanup is placed on companies or individuals determined to be responsible for contamination.

EPA staffs a toll-free number to answer CERCLA-related questions. This number is 1-800-424-9346, and is the same phone number for questions about hazardous waste regulations.

Background

The other regulations discussed in this book have a prospective focus: They deal with future and existing activities, not past occurrences. They are not intended to provide remedies for situations where environmental damage has already occurred and responsible parties are unidentifiable, unavailable, or lacking in funds. CERCLA, also known as Superfund, was enacted to address such situations.

The impetus for CERCLA largely consisted of the 1978 discovery of hazardous chemicals seeping into homes at Love Canal, New York, and a desire to address this and similar situations. The resulting legislation was admittedly a compromise bill, enacted literally during the last hours of the 96th Congress. The final law, signed by President Carter on December 11, 1980, eliminated many of the provisions in earlier drafts and limited funding mechanisms to a life of five years unless reauthorized. Many legislators felt such measures were acceptable in the face of the perceived urgent need for some type of legislation. Fittingly, some of the first Superfund expenditures were for the cleanup of Love Canal.

Although reauthorization legislation proved to be extremely controversial, Congress approved a five-year Superfund extension late in 1986. This reauthorization bill provides $9 billion over a five-year period to clean up abandoned hazardous waste sites.

Content

CERCLA grants authority for EPA to provide required remedial actions whenever any hazardous substance is released into the environment or there is a substantial threat of such a release. The scope of this authority is vast, since the key terms contained in the act have extremely broad statutory definitions. For example, hazardous substance includes any material so designated by the Clean Water Act (CWA), the Clean Air Act (CAA), the Toxic Substances Control Act (TSCA), or the Resource Conservation and Recovery Act (RCRA). Under section 102 of CERCLA, substances may be added, which will be included in the definition. To release includes spilling, leaking, pumping, pouring, emitting, emptying, discharging, injecting, escaping, leaching, dumping, or disposing. The environment is defined as navigable waters, any other surface waters, groundwater, drinking water supplies, land surface or subsurface strata, and ambient air.

The major provisions of CERCLA address the following: response actions, liability, financing response actions, and notification requirements. Amendments adopted in 1986 added significant provisions which address emergency planning at state and local levels.

CERCLA authorizes EPA to dispose of removed material, conduct monitoring programs, limit site access, evacuate affected individuals, and undertake studies, including monitoring, testing, and surveys. Types of response actions are summarized in Table 6.6, and a schematic representation of a theoretical Superfund response is shown in Figure 6.6. The proce-

Table 6.6
Types of Response Actions

Removals - Short term responses to give immediate relief from significant dangers:

- Limited to six months and one million dollars
- Immediate removals - emergency situations requiring an immediate response to a site posing immediate and significant danger
- Planned removals - situations requiring an expedited but not immediate response

Remedial actions - Long term projects to provide permanent remedies to National Priorities List (NPL) sites:

- NPL designation
- Remedial investigation (RI)
- Feasibility study (FS)
- Record of decision (ROD)
- Design and construction
- Operation and maintenance
- Cost recovery

Figure 6.6
Superfund Response

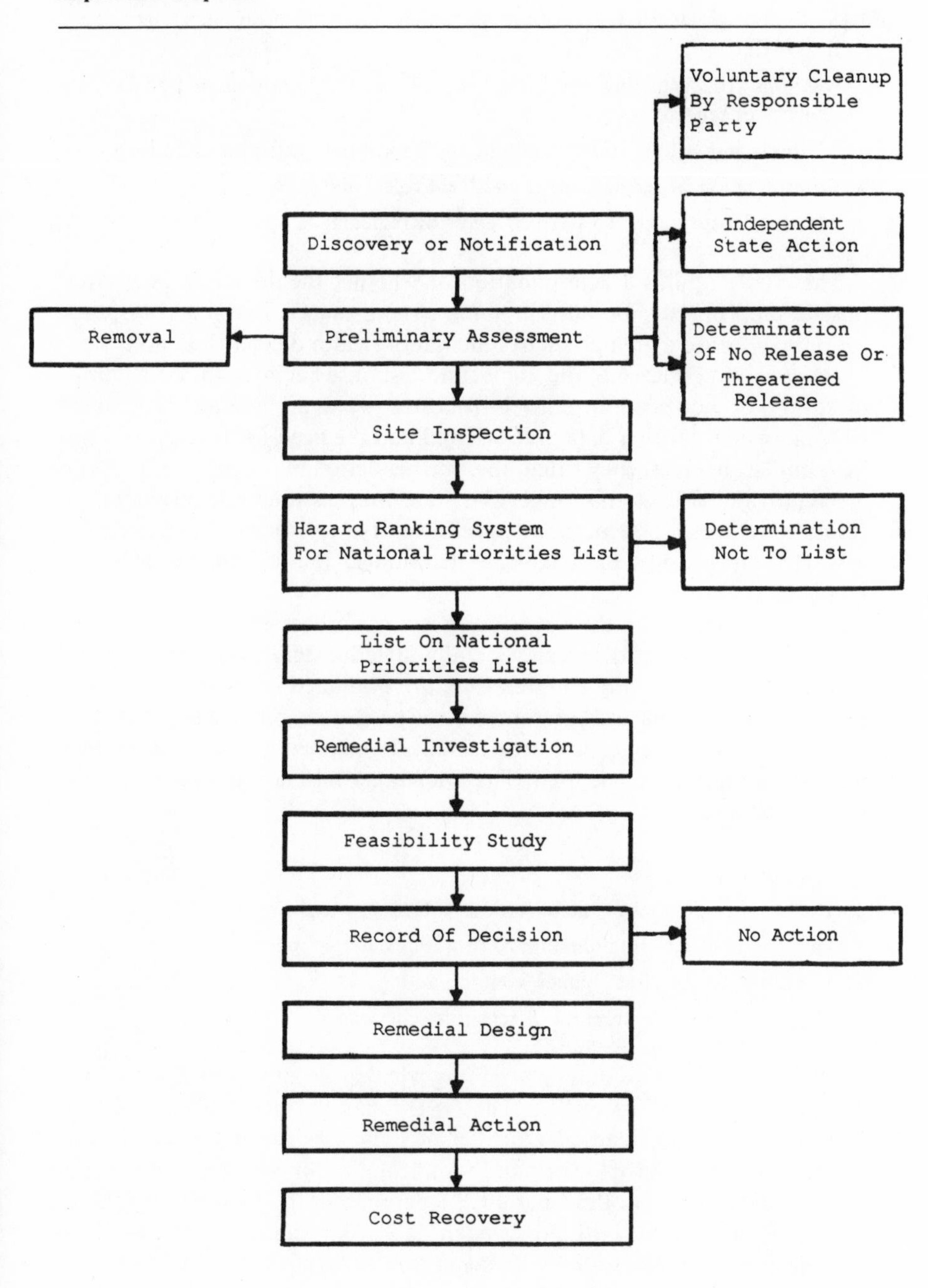

dures comprising this response are defined in the NCP, which was first required by the CWA and includes

1. Methods for identifying facilities at which hazardous substances have been disposed;
2. Methods for evaluating and remedying releases of hazardous substances and analysis of relative cost;
3. Methods and criteria for determining the appropriate extent of cleanup;
4. Determination of federal, state, and local roles; and
5. Means of assuring the cost-effectiveness of remedial actions.

The NCP requires a determination of whether the no-action alternative may be appropriate, because in some circumstances a remedial response action may cause a greater environmental or health danger than no action.

As shown in Figure 6.6, the Superfund response begins with some type of discovery or notification. This is followed by a preliminary assessment (PA), in which existing data are reviewed and the need for further action is determined. In emergency situations, further action may consist of immediate removal. Also at this stage, the state may assume responsibility for resolution of the problem, the responsible party may be persuaded to undertake cleanup actions, or it may be determined that no further action is required.

If the site remains in the Superfund response process, the PA is followed by a site inspection (SI), which generally includes significant sampling and analytical work. Existing and new data are evaluated in an effort to determine the potential hazards presented by a specific site relative to other sites. In an attempt to apply uniform technical judgment, a specific Hazard Ranking System (HRS) is used. The HRS takes into account the following considerations:

1. The population at risk;
2. The hazard potential of the hazardous substances involved;
3. The potential for contamination of drinking water supplies;
4. The potential for direct human contact; and
5. The potential for destruction of sensitive ecosystems.

These factors are used in determining whether a site will be listed on the National Priorities List (NPL). The NPL ranks the sites which appear to warrent Superfund-financed remedial actions. The list is revised at least once a year. States have the primary responsibility for selecting and evaluating candidate sites, but the federal EPA determines the final NPL ranking. Sites designated on the NPL are listed in Tables A.10 and A.11 in the appendix. Remedial actions can only be taken at sites identified on the NPL.

A remedial investigation (RI) is performed for sites which are placed on the NPL. During an RI, there is further evaluation of existing data and definition of required additional data. The site characterization yields a determination of appropriate initial response actions.

The RI is followed by a feasibility study (FS) in which alternative remedies are evaluated. At this time, as required by the NCP, there must be a determination of whether the no-action alternative is appropriate. In some circumstances, a remedial response action may cause a greater environmental or health danger than no action.

The record of decision (ROD) documents the decision-making process for selecting the chosen action. During the ROD, the basis for the selected level of cleanup is documented and justified. If a remedial response is selected, the ROD is followed by design and implementation of the remedial action and by cost-recovery efforts.

CERCLA establishes strict liability (i.e., liability without fault) for damages resulting from hazardous wastes specified under the act. Once government funds are used for site cleanup, the government has the authority to sue the responsible party independent of a determination of negligence. There are three classes of potential responsible parties: owners and operators, generators, and transporters.

Liability extends to all removal and remedial costs incurred by federal or state governments consistent with the NCP, other necessary response costs incurred consistent with the NCP, and damages for injury to natural resources under the jurisdiction of a governmental entity. The original CERCLA imposed no liability for federally permitted releases or for personal injury resulting from releases of hazardous substances. Under the 1986 amendments, citizen suits are permitted where the federal government has failed to perform any nondiscretionary duty or act or is in violation of the Superfund law.

Title II of CERCLA provided monies to finance response actions through a Hazardous Substance Response Trust Fund, funded by taxes on crude oil, certain petroleum products, chemical feedstocks, and appropriations from general revenues. This fund came to be known as "Superfund." Collection of the taxes commenced in April 1981 under the original act.

The 1986 reauthorization legislation includes clauses which call for a collection of $1.4 billion from taxes on chemical feedstocks, $2.75 billion from a two-tiered tax on crude oil, $2.5 billion from a new corporate income tax, and $600 million from interest and cost recoveries. A separate $500 million trust fund for leaking underground storage tanks will be funded by a 0.1 cent per gallon tax on motor fuels. Although once proposed, there is no tax on hazardous waste.

The notification requirements of CERCLA address both spills and disposal sites. Sections 102 and 103 require the reporting of the release of hazardous substances into the environment, in cases where the release is not allowed in accordance with a permit. A federally permitted release is any discharge or emission authorized by permits issued under the CWA, Safe Drinking Water Act (SDWA), CAA, RCRA, Marine Protection, Research, and Sanctuaries Act (ocean dumping), or the Atomic Energy Act (AEA). All persons presently or previously owning facilities at which hazardous substances were handled, and all transporters to such sites, are required to notify EPA of their actions if releases occur or are determined to have occurred.

The amount of released substance which triggers CERCLA notification requirements is the reportable quantity (RQ) as defined in the CWA. If a specific RQ for a substance has not been otherwise established, CERCLA defines the RQ as one pound. No notification is required for

1. Releases that have been reported to the National Response Center (NRC) under Subtitle C of RCRA;
2. The application of registered pesticides in accordance with EPA requirements; or
3. Stable, continuing releases that have been previously reported, or issued from a previously reported site.

Title III of the 1986 Superfund Amendments and Reauthorization Act (SARA) contained new emergency planning and response requirements, known as the Emergency Planning and Community Right-to-Know Act of 1986 (EPCRA). EPCRA addresses the following:

1. Emergency planning notification;
2. Emergency release notification;
3. Reporting on chemicals and releases for community right-to-know organizations.

EPCRA does not override state and local laws, but it does require states to establish a state emergency response commission, which in turn must designate local emergency planning districts and committees.

In response to other requirements contained in EPCRA, EPA compiled and finalized an extensive list of toxic chemicals in April 1987, which is shown in Table A.12. The list identifies 406 extremely hazardous substances, and associated quantities at which they have been determined to pose health risks. This quantity is termed the threshold planning quantity (TPQ). Anyone who stores, uses, manufactures, or ships any of these substances in greater quantities than the TPQ must comply with EPCRA requirements.

Other provisions in SARA include clean-up standards, higher penalties, requirements for toxicological studies of common pollutants, and a schedule for EPA to begin action at sites on the NPL. The schedule section requires preliminary assessments to be completed for all identified potential Superfund sites and SIs to be completed by January 1, 1989. Remedial investigations and feasibility studies are to be initiated at 275 NPL sites within three years, and remedial action is to be initiated at 375 sites within five years.

Under SARA mandates, EPA released a second list of chemicals, shown in Table A.13. This list is composed of 100 substances found at Superfund sites, selected from an original list of 717 substances identified on the basis of chemical toxicity, frequency of occurrence, and potential for human exposure. These 100 substances have been prioritized for further study by EPA and will therefore likely be some of the first chemicals addressed by future CERCLA regulations.

Enforcement

CERCLA is enforced by EPA. CERCLA-mandated clean ups may be coordinated with state and local governments. A state's participation in Superfund implementation is predicated on that state's ability to assume responsibility for clean-up initiatives.

The tasks that the state may assume include site inventory, ranking, analysis, feasibility study and development of clean-up alternatives, selection of contractors and subcontractors, management of clean-up work, provision of 10 percent match for construction, and operation and maintenance work.

The selection of the clean-up alternative financed by funds provided under CERCLA must be a federal decision.

Relationship to Other Regulations

CERCLA required revisions in the NCP, which was first drafted under the CWA. The minimum quantity requiring notification of spills is one pound, unless EPA regulations in CERCLA or section 311 of the CWA specify a greater quantity. Treatment, transportation, and disposal of hazardous wastes from sites regulated by CERCLA must meet standards in RCRA. Other clean-up criteria may be based on standards contained in the CWA or SDWA. Any substance designated as hazardous under CERCLA must bear the same designation under the HMTA. CERCLA provisions function in conjunction with RCRA provisions regarding leaking underground storage tanks, and in conjunction with OSHA provisions regarding right-to-know programs. Additionally, SARA required OSHA to issue regulations to protect Superfund clean-up workers. SARA resulted in amendments to both the SDWA and RCRA.

Potential Future Developments

Future significant legislative amendments to CERCLA are unlikely in the near future, since major reauthorization legislation (SARA) was passed in 1986. Due to schedule-related provisions in the amendments, the Superfund program and related development of regulations are expected to move ahead at an accelerated pace.

Before the passage of SARA, hazardous waste sites on federal lands were the sole responsibility of the facility that caused the problem, and therefore were not subject to CERCLA regulations. Under SARA, EPA has jurisdiction over federal sites, and its general cleanup standards must be followed. In 1987, 32 federal sites were added to the NPL, and over 200 are reported to be targeted for the NPL evaluation.

Legal issues involving CERCLA are expected to remain numerous and active, especially since the authority granted to the federal government for remedial actions under CERCLA far surpasses the imminent hazard provisions of RCRA and other environmental statutes. Therefore, CERCLA is the logical first-choice vehicle for federal suits involving imminent hazards to the environment.

CERCLA currently provides no funding for an oil spill liability and compensation program. Such provisions were dropped from the House version of the Superfund bill in return for a Senate commitment to consider an oil spill program as a separate piece of legislation. Whether any such future legislation would be attached to CERCLA is questionable.

7

ASBESTOS CONTROL REGULATIONS

If we can't regulate asbestos, we can't regulate anything.

—John Moore, EPA
Assistant Administrator for Pesticides and Toxic Substances

OVERVIEW

Asbestos is the name for a group of naturally occurring fibrous mineral substances which have been widely used in the building industry because of their fire-resistant and insular properties. When potential health hazards became associated with exposure, asbestos came under concentrated governmental scrutiny in the 1970s, resulting in widespread concern with the abatement of existing asbestos containing materials (ACM).

The health effects of asbestos exposure are exhibited after respiratory exposure has taken place, with asbestos-related diseases occurring because people breathe airborne particles or fibers of the mineral. Asbestos tends to break up into fibers of microscopic size, which can remain suspended in the air for long periods of time, and can easily penetrate body tissues when inhaled. Because of their durability, these fibers can remain in the body for many years.

Because only respiratory exposure to asbestos has been documented as disease-causing, pertinent regulations place emphasis on the physical state of ACM. The condition of most concern is termed "friable": capable of being crumbled, pulverized, or reduced to powder by hand pressure.

Asbestos in buildings is addressed by specific sections of three of the regulations discussed in preceding chapters: the Occupational Safety and Health Act (OSH Act), Clean Air Act (CAA), and the Toxic Substances Control Act (TSCA). It is also the subject of the Asbestos School Hazard Abatement Act (ASHAA) and the Asbestos Hazard Emergency Response Act of 1986 (AHERA).

EPA staffs an information number to answer questions regarding asbestos. Names of laboratories that are qualified to test and analyze asbestos samples and certain technical assistance can be obtained by calling 202-554-1404.

BACKGROUND

Construction materials containing asbestos have been used extensively in schools and other buildings, in such widely varying applications as structural fireproofing, accoustical materials, roofing felt, vinyl floor tiles and flooring, patching compounds, textured paints, ceilings, stoves, furnaces, wall and pipe insulations, roofing, shingles, and siding. Estimates of the quantity of asbestos used in various products from 1965 to 1978 range from 2.2 million to 5 million tons. Table A.14 in the appendix lists ACM commonly found in buildings.

Concern about asbestos exposure was initially based on data linking various respiratory diseases with exposure in the shipbuilding, mining, milling, and fabricating industries. Most incidents of asbestos-related diseases (asbestosis, lung cancer, and mesothelioma) have resulted from exposure to high levels of asbestos in such asbestos-related industries prior to 1972. Although a definitive correlation between asbestos exposure and disease has been established, it is predicted that only a small proportion of people exposed to low levels of asbestos (non-asbestos industry-related) will develop asbestos-related diseases. Smokers, children, and young adults are at a greater risk than the general population.

Soon after definitive demonstration of adverse health effects, control regulations began to be promulgated. The Occupational Safety and Health Administration (OSHA) issued its first standard for occupational exposure to asbestos in 1971, followed by a more stringent rule initially issued as an emergency temporary standard in 1972. In 1975, a revision was proposed to make the standards apply to all work environments covered by the OSH Act, excluding the construction industry, for which a separate revision was applicable. OSHA lowered the permissible exposure level by an order of magnitude (from 2 to 0.2 fibers per cubic centimeter) in June 1986.

REGULATION OF ASBESTOS IN BUILDINGS

Since production of asbestos-containing products has become severely limited since 1973 (decreasing more than 75 percent), this discussion will address the most common source of asbestos exposure: asbestos in buildings.

Resulting asbestos exposure is often controlled through abatement actions, which in themselves necessitate protective standards for workers conducting abatement projects. The need for abatement is determined on a case-by-case basis, normally initiated by a survey for ACM, as shown in

Figure 7.1. Abatement actions include removal, encapsulation, and enclosure. Details about the applicability of these abatement options are presented in Table A.15 in the appendix.

OSHA exposure standards and the portions of EPA regulations governing the handling and removal of asbestos apply to abatement actions. Additionally, asbestos removal contractors are encouraged to employ safety procedures beyond the minimum regulatory requirements of EPA and OSHA. To this end, EPA publishes various guidance documents dealing with asbestos abatement. These documents contain valuable information and recommendations which are not enforceable as regulations.

Initially, no federal law required abatement activities, but numerous abatement programs were undertaken because of local regulations or exposure concern. Among the directives contained in AHERA was a requirement for EPA to promulgate regulations pertaining to situations requiring response actions, and the implementation of these actions. EPA began to propose these regulations in 1987.

Both EPA and OSHA have published regulations intended to reduce asbestos exposure resulting from asbestos in buildings, mainly addressing active abatement situations. The majority of these regulations are contained in the CAA, TSCA, and OSHA standards.

Emissions of asbestos to the ambient air are controlled under section 112 of the CAA, which establishes the National Emission Standards for Hazardous Air Pollutants (NESHAPs). The regulations specify control requirements for asbestos emissions, including work practices to minimize the release of asbestos fibers during handling of asbestos waste materials. The following requirements are specifically applicable to ACM abatement activities:

1. When a building is demolished (or more than 260 linear feet of asbestos surfacing material are removed during renovation), advance notice must be filed with the EPA regional office and/or the state, giving name and address of the building owner or manager; description and location of the building; scheduled starting and completion dates of ACM; description of the planned removal methods; and name, address, and location of disposal site.
2. ACM can be removed only with wet removal techniques unless dry removal is allowed with written EPA approval due to special conditions.
3. No visible emissions of dust are allowed during removal, transportation, and disposal of ACM.

EPA implemented a separate asbestos control regulation under TSCA in 1984: "The Friable Asbestos-Containing Materials in Schools; Identification and Notification Rule," known as the Asbestos-in-Schools Rule. This regulation requires that all secondary and primary schools (public and private) be inspected to determine the presence and quantity of asbestos, that the local community be notified, and that the building be posted. No

Figure 7.1
Initial Steps in an ACM Survey

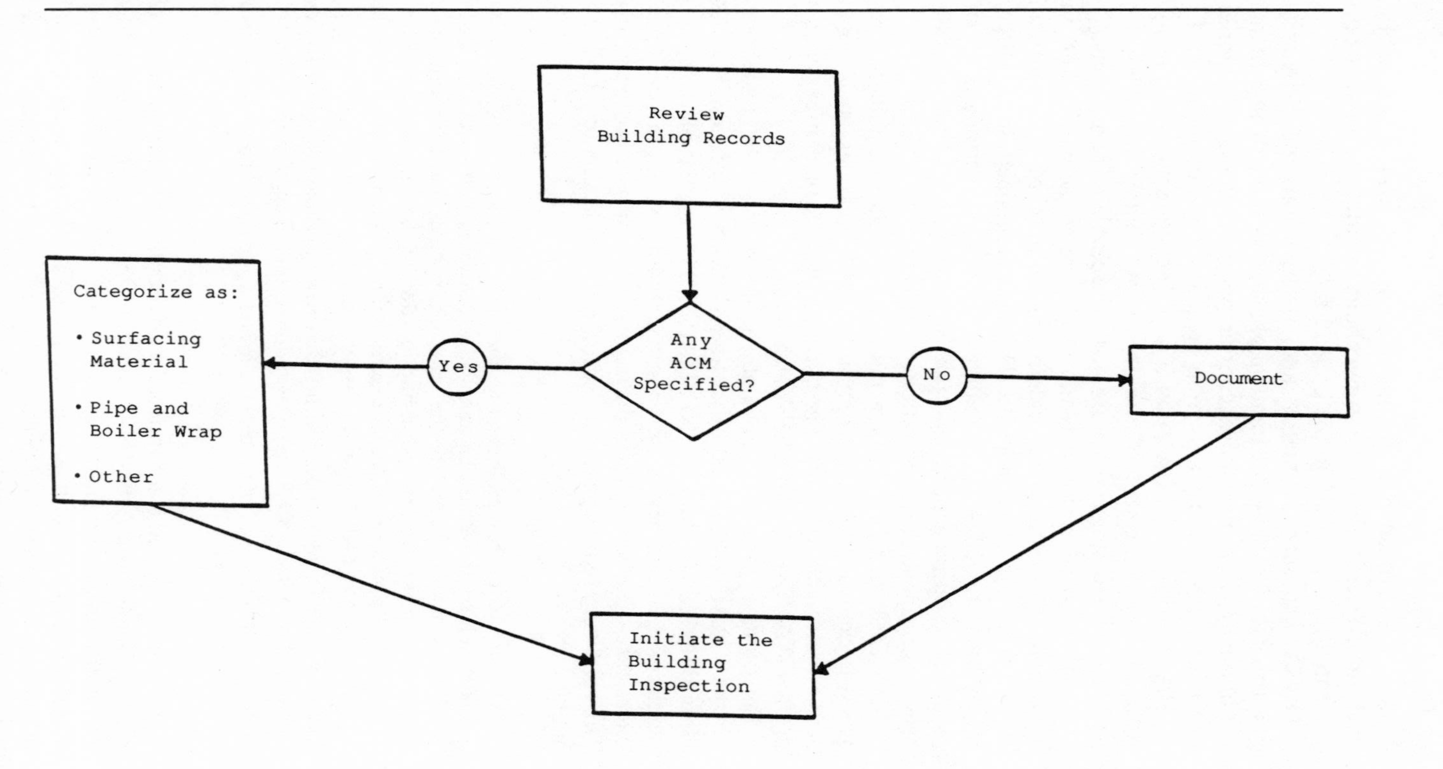

parallel rule currently applies to other public or commercial buildings. Corrective actions, such as asbestos removal or encapsulation, are currently left to the discretion of the school administrators. Also, under section 8(a) of TSCA, EPA has issued a final rule to require reporting of production and exposure data concerning asbestos.

Although not a part of the Asbestos-in-Schools Rule, ASHAA is related to this rule, and some portions of both are administered jointly at EPA. ASHAA established a $600 million grant and loan program to assist schools with asbestos abatement projects. The program also provided for the compilation and distribution of information concerning asbestos, and the establishment of standards for abatement projects and abatement contractors.

AHERA was enacted January 1986 and is the legal vehicle under which federal abatement requirements will be promulgated. The act requires formulation of regulations under TSCA to address asbestos hazards in schools, requiring all public and private elementary and secondary schools to conduct inspections for ACM and to develop management plans describing abatement actions which will be undertaken. Schools will be required to implement their management plans and complete them in a timely fashion. The act additionally requires states, using an EPA model, to develop accreditation programs for workers who inspect for asbestos, develop management plans, and conduct abatement work.

OSHA regulations specify airborne exposure standards for workers, engineering and administrative controls, workplace practices, and surveillance and worker protection requirements. These regulations apply to all workplace activities involving asbestos, including the removal of ACM from buildings.

Initially the OSHA regulations specified a maximum workplace airborne asbestos concentration of two fibers per cubic centimeter (cc) on an eight-hour time-weighted average basis. OSHA has subsequently decreased this exposure limit by an order of magnitude to 0.2 fibers/cc. Respirators are not to be used to achieve the permissible exposure levels under normal working conditions. They are allowed as a compliance device during the time required to install controls or implement work practice controls designed to reduce exposure to or below the permitted exposure limits, which applies to abatement activities.

Every employer must provide medical examinations for employees engaged in occupations which involve exposure to airborne asbestos fibers. These include pre-placement, annual, and termination examinations. These exams generally consist of at least a chest x-ray, a medical history or study to determine the presence of any possible respiratory diseases, and a pulmonary function test. Records of these exams must be retained a minimum of 30 years to provide documentation of the health status of the employee.

DISPOSAL REGULATIONS

Asbestos-containing wastes are not hazardous as defined by the Resource Conservation and Recovery Act (RCRA), although states can enact more restrictive regulations which may define such wastes as hazardous. Obviously, such classification greatly impacts the transportation and disposal of the waste, so local agencies must be consulted.

EPA rules governing the disposal of ACM wastes consist of requirements for active and inactive disposal sites under NESHAPs and specify general requirements for solid waste disposal under RCRA. Advance EPA notification of the intended disposal site is required by NESHAPs. It is further recommended that the containerized waste be covered within 24 hours with a minimum of six inches of nonasbestos material. Improperly containerized waste is a violation of the NESHAPs.

Both EPA and OSHA regulations require the wastes to be containerized as necessary to avoid creating dust during transport and disposal. The generally recommended containers are 6-mil-thick plastic bags, sealed in such a way as to make them leak-tight. Asbestos waste slurries can be packaged in leak-tight drums if they are too heavy for the plastic bag container. Both EPA and OSHA specify that the containers be tagged with a warning label. Either the EPA or OSHA label may be used.

ENFORCEMENT AND RELATIONSHIP TO OTHER REGULATIONS

The various regulations discussed in this section are enforced by the agencies responsible for the overall program under which the specific regulation falls. Thus, the Department of Labor (DOL) enforces occupational safety and health regulations promulgated under the OSH Act, and EPA enforces the asbestos-related portions of the CAA and TSCA.

States and local agencies may have more stringent requirements than the federal standards. As of 1987, 32 states had enacted more than 60 asbestos-related laws.

Other federal legislation regulating asbestos includes the Clean Water Act (CWA), under which EPA has set standards for asbestos levels in effluents to navigable waters; the Mine Safety and Health Administration (MSHA), which oversees the safety of workers involved in the mining of asbestos; the Consumer Product Safety Commission (CPSC), which regulates asbestos in consumer products; the Food and Drug Administration (FDA), which is responsible for preventing asbestos contamination in food, drugs, and cosmetics; and the Department of Transportation (DOT) which regulates transportation of the mineral and ACM.

POTENTIAL FUTURE DEVELOPMENTS

Legislation signed in 1986 established a national accreditation program for contractors who clean up friable asbestos in schools. Specific certification and qualification programs are being developed by individual states. In April of 1987, EPA announced far-reaching draft regulations that would require every school board in the nation to take action regarding ACM, even if that action is only inspection.

Potential future regulations are likely to address assessments of asbestos in public buildings other than schools, funding mechanisms for such assessments, requirements for abatement actions, resolution of liability issues, and a potential ban of all new asbestos containing products (originally proposed by EPA in January 1986). All such measures are extremely controversial, in spite of the fact that EPA's scientific case against asbestos is probably the strongest in the long list of toxic substances that the agency is supposed to regulate. The controversy surrounding further regulation of asbestos results from the high costs involved in abatement work and the inherent scientific and political uncertainties involved in assessing, evaluating, and managing public health risks.

EPA is scheduled to revise the CAA NESHAPs by late 1988. Revisions may result in more stringent requirements and additional reporting procedures.

EPA has listed asbestos as a covered substance under the federal Superfund law, but this action has been overturned by a federal district court. Final resolution of this matter will have far-reaching economic impacts.

Any ban on asbestos manufacturing, importation, and processing will probably include a permit/exemption system to facilitate a gradual phase-out of asbestos products. In 1987 the American Society of Mechanical Engineers (ASME) advised EPA, among other things, that there were insufficient data on the safety of alternative materials to set a deadline for removing asbestos from all auto and truck brakes.

8

RADIOACTIVE MATERIALS CONTROL REGULATIONS

> Nature is neutral. Man has wrested from nature the power to make the world a desert or to make the deserts bloom. There is no evil in the atom; only in men's souls.
>
> —Adlai Stevenson, 1952

OVERVIEW

Federal agencies involved in radioactive materials-related activities may be divided into three categories: those that use radioactive materials, those that perform and sponsor research, and those that safeguard health and the environment. The possession, use, transfer, storage, and disposal of radioactive materials within the United States is regulated. This regulation is performed by federal agencies such as EPA, the Nuclear Regulatory Commission (NRC), the Department of Transportation (DOT), the Department of Energy (DOE), the Food and Drug Administration (FDA), and the Occupational Safety and Health Administration (OSHA); and Agreement States, which regulate most radioactive materials in that state in lieu of a federal agency.

The NRC is the dominant agency and historically has set the nuclear industry's operating guidelines and rules. However, each involved agency retains authority over specific aspects associated with radioactive materials. Many of the regulations pertaining to radioactive materials have evolved to the point where the overlaps and inconsistencies have been minimized, but overlapping agency jurisdiction remains an unavoidable aspect of radioactive materials control.

The following discussion will focus on agency activities with the primary function of environmental and public health protection and the regulations intended to carry out this function.

BACKGROUND

Effects of radiation on laboratory animals were reported as early as 1901, and an initial attempt to define acceptable exposure conditions for occupational exposure to x-rays occurred in 1902. The first formal statement of recommended radiation exposure protection measures was adopted in the United States in 1922, and subsequently the awareness of the need for protection standards increased significantly. Today, various agencies regulate specific aspects of radioactive materials, authorized by several legislative acts.

The Atomic Energy Act of 1954 (AEA), as amended, created the Atomic Energy Commission (AEC). The AEC was responsible for conducting research and development on materials and facilities for all uses of atomic energy. In order to separate regulatory functions from promotional activities, the AEC was divided in 1974 into the NRC and the Energy Research and Development Administration (ERDA). The functions of ERDA were transferred to DOE in 1977. Although basically a user agency involved in the production of nuclear weapons and development of radioisotopes, DOE has some regulatory functions through standards setting and determination of contractor requirements. As reflected in its name, the primary function of the NRC is a regulatory one.

In the late 1950s, there was concern that the majority of radiation protection standards were issued by private organizations rather than by the federal government. In response to this concern, President Eisenhower created the Federal Radiation Council (FRC) in 1959. The council was to advise the President concerning radiation matters affecting health and give guidance concerning the formulation of radiation protection standards.

EPA's primary radiation responsibility was transferred to it when the agency was formed in 1970, and the FRC was subsequently abolished. As authorized by the AEA, EPA is directed to establish regulations intended to protect the environment from radioactive materials. In related activities, OSHA is authorized to establish regulations intended to protect workers from occupational exposure to radiation, and the FDA regulates the medical efficacy of radioactive drugs and devices through its Bureau of Radiological Health (BRH).

The transportation of radioactive materials has been regulated since the early 1950s, initially resulting from a concern about the fogging of photographic film transported in the proximity of radioactive material. The increased complexity of regulating transportation hazards due to radioactive materials necessitated extensive regulatory changes in 1983.

Congress historically has found it difficult to agree on a comprehensive radioactive waste management policy. In December 1980, Congress enacted the Low-Level Radioactive Waste Policy Act, making states more responsible for the disposal of low-level radioactive waste. In 1982, Congress enacted the Nuclear Waste Policy Act, which kept the responsibility for high-level radioactive waste disposal with the federal government.

USAGE REGULATIONS

The NRC has retained regulatory control in areas that include weapons-grade nuclear material, international shipments, and nuclear power plant operation. Some states which have elected to administer the regulations set by the NRC are known as Agreement States. An Agreement State is a state that has assumed regulatory authority from the AEC or the NRC over radioactive by-products, source materials, and small quantities of special nuclear materials. In 1985 the Agreement States included Alabama, Arizona, Arkansas, California, Colorado, Florida, Georgia, Idaho, Kansas, Kentucky, Louisiana, Maryland, Mississippi, Nebraska, Nevada, New Hampshire, New Mexico, New York, North Carolina, North Dakota, Oregon, Rhode Island, South Carolina, Tennessee, Texas, and Washington. The NRC audits the performance of Agreement States to assure compatibility with federal standards and can override a state's authority if sufficient cause is demonstrated.

Agreement States issue general and specific licenses as does the NRC. General licenses are issued to facilities handling low or inconsequential quantities of radioactive materials. Specific licenses are issued for certain defined isotopes and quantities used under specific criteria. A broad license allows for more on-site flexibility than either a general or specific license and is usually administered by a committee. The NRC and Agreement States also regulate the components of radiation protection plans, including personnel monitoring, posting of areas, training, and maintenance of records and inventories.

For the most part, Agreement States have adopted rules compatible with the federal rules. Slight variations exist, and, in some states such as California, New York, New Jersey, and New Hampshire, significant changes are in place. Consequently, any use of radioactive materials in a particular location should be preceded with a review of the specific local regulations.

Historically, EPA has been involved in promulgating usage-related regulations dealing with x-rays, the uranium fuel cycle, and naturally occurring radioactive materials. Many of these regulatory efforts were not successful due to controversies surrounding overlapping agency jurisdictions. EPA and the now-defunct Health, Education, and Welfare (HEW) Department jointly issued a guidance document in 1978 addressing radiation protection for diagnostic x-rays in federal facilities, but EPA is not currently involved in the enforcement of x-ray control regulations. EPA's attempts to regulate the uranium fuel cycle were protested by the AEC, with the result of delaying and weakening the entire standards-setting effort.

EPA's current regulations for radioactive materials pertain mainly to environmental contaminants from natural sources such as uranium mining and phosphate processing and are found in 40 CFR 190 Subchapter F. Additionally, the Clean Air Act (CAA) and Safe Drinking Water Act (SDWA) regulate radioactive materials. Reportable quantities (RQs) for radionuclides are set by EPA in accordance with the Clean Water Act (CWA) and the Comprehensive Environmental Response, Compensation, and Liability Act (CERCLA). EPA is also responsible for advising the President with respect to environmental radiation matters affecting health.

The FDA is authorized under several statutes to regulate electronic products, drugs, and medical devices that emit radiation. Most of these activities are performed by the BRH.

In coordination with NRC regulations, OSHA regulates the level of radioactivity to which workers may be exposed. This has the effect of indirectly defining certain usage practices.

TRANSPORTATION REGULATIONS

DOT has established the rules for proper packaging, description, labeling, placarding, manifesting, and transport of radioactive materials. These regulations have mainly evolved in accordance with international transport regulations as promulgated by the International Atomic Energy Agency (IAEA). In some instances, the NRC, the Agreement States, or local agencies enforce additional rules.

Radioactive materials regulated for purposes of transportation are defined as those which spontaneously emit ionizing radiation and which have a specific activity exceeding 2μCi/kg. Materials with lower specific activities are not subject to transportation regulations due to their radioactivity. Regulated radioactive materials are divided into the following categories for transportation purposes:

- Excepted quantities, very low total activity;
- Limited quantities, limited specific activity;
- Low Specific Activity (LSA), limited specific activities, but greater than limited quantities;
- Type A quantities, moderate activities;
- Type B quantities, larger activities;
- Highway Route Controlled Quantities (HRCQ), very large amounts under very controlled packaging, transport, and routing; and
- Fissile materials, those capable of undergoing spontaneous nuclear fissioning.

Each category has specific requirements intended to provide suitable containment of the material, protection from radiation exposure, rejection of decay heat, and prevention of criticality. The stringency of the requirements depends on the determined hazard of the material being transported.

Labeling and placarding requirements are also regulated by DOT, as is the radiation shipping record, which is a detailed manifest-type document that completely describes the isotope, quantity, packaging, labeling, placarding, exposure rates, chemical and physical characteristics, licenses of destination and shipper, and emergency numbers.

Various modes of shipment include rail, ship, air, and land vehicle. A separate set of specific rules is associated with each of these shipment modes. Additional agencies may also be involved depending on the method of transport.

WASTE MANAGEMENT REGULATIONS

Radioactive waste can be divided into three broad categories: high-level waste (HLW), transuranic waste (TRU), and low-level waste (LLW).

HLW is the liquid or solid waste resulting from the chemical reprocessing of irradiated (spent) reactor fuel, or other highly radioactive materials determined by the NRC to require permanent isolation. Because of a 1977 decision to prohibit reprocessing of spent fuel, only a small quantity of commercial high-level waste (CHLW) exists today. About 97 percent of the volume of HLW was generated by government defense-related activities, and is known as defense high-level waste (DHLW). DHLW is stored in tanks at military reservations until facilities are available to treat and dispose of these wastes. The federal government is responsible for the safe disposal of DHLW as authorized by the Nuclear Waste Policy Act of 1982. Within the federal government, DOE has the responsibility for identifying disposal sites, and NRC has the responsibility for establishing disposal guidelines consistent with EPA standards.

Only small quantities of commercial TRU currently exist, mainly originating from research projects. Much of this waste is in the form of contaminated combustible materials, so incineration and acid decomposition are often effective means of reducing waste volume to facilitate disposal. NRC regulates these processes when applied to TRU. Defense-related TRU wastes are currently being stored on government reservations.

LLW is defined primarily on the basis of what it is not, rather than what it is: radioactive waste not classified as high-level radioactive wastes, or spent nuclear material. LLW constitutes over 80 percent of the volume of all nuclear wastes, but only about 2 percent of the total radioactivity is contained in these wastes. Not all LLW requires disposal in shallow land disposal sites. The NRC regulations allow the release to the environment of waste liquids and gases which contain specified low concentrations of radionuclides. In addition, if the waste contains radionuclides with very short half-lives, it may be stored until the isotopes decay to insignificant levels and then disposed of as normal trash.

The Low-Level Radioactive Waste Policy Act of 1980 made states responsible for the management of LLW. Regions have been established around groups of states which have been charged with the creation of a compact. A compact is a legally binding interstate agreement that requires enactment by the respective state legislature and governors. Compacts must identify and establish a disposal facility and have mechanisms in place to manage the waste generated by those states. Functioning compact states are authorized, upon congressional consent, to exclude wastes from outside their compact region. Requirements for establishing and operating commercial LLW disposal facilities are defined by the NRC and the DOE.

EPA, the NRC, and maritime regulations may grant permits for ocean disposal of LLW, but stringent requirements have prevented receipt of such permit applications to date. Ocean disposal permits may be issued only when no alternative exists, and require both an impact assessment and congressional approval.

ENFORCEMENT AND RELATIONSHIP TO OTHER REGULATIONS

The control of radioactive materials is accomplished through a compendium of assorted regulations. Complete enforcement authority for radioactive materials control does not rest with one single agency but is the responsibility of the agency administering the program under which a specific regulation has been promulgated. It is important, therefore, to be aware of the enormous diversity of agencies that may become involved in usage, licensing, transportation, disposal, spills, overexposures, and the like. Each agency has its sphere of control, which may overlap with that of another agency.

Depending on the specific regulation, final enforcement may be the responsibility of either a federal or state agency. The major federal enforcement agencies for environmentally related regulations include EPA, DOT, and the NRC. OSHA is the federal enforcement agency for occupational regulations pertaining to radioactive materials.

POTENTIAL FUTURE DEVELOPMENTS

Usage regulatory developments continue to be concentrated in the areas of expanding regulatory programs to provide increased personal and environmental protection, while resolving issues of overlapping jurisdictions. DOT regulations are expected to continue to conform to IAEA standards, accomplished by regulatory amendments to adopt new IAEA rules as they are published.

In response to increasing concern over radon levels in homes, EPA is involved in a radon research program and may promulgate standards addressing this problem in the future. EPA is also taking a major role in regulating mixed wastes. These are wastes with hazardous characteristics in addition to radioactivity. EPA is expected to increase its involvement in other aspects of radioactive materials control, but efforts in this area may be diluted and delayed by other regulatory priorities.

Disposal of both HLW and LLW remains a controversial issue, with emphasis being concentrated on the location of disposal sites.

In 1986, DOE named sites in Nevada, Texas, and Washington State as finalists for HLW disposal. Critics of the site selection accused the DOE of basing its decision on politics, rather than on scientific criteria. In early 1987, the DOE announced a five-year delay in opening the first repository for the HLW and postponed until the mid- to late-1990s selection of final candidates for a second site. In the meantime, the DOE plans to proceed with plans for a short-term storage and waste handling site known as a Monitored Retrievable Storage (MRS) facility.

Controversy concerning LLW disposal continues to focus on the possibility of states in nonfunctioning compacts having out-of-state disposal sites closed to them, in addition to intra-compact efforts to resolve siting dilemmas.

Technological developments which do not require land disposal of LLW are one area of concentrated research with both the DOE and industry. This effort has related regulatory standard development and associated political controversies.

9

PROGRAM MANAGEMENT

> Compromises must be reached between the ultimate goals of legislation and the realistic ability of institutions to perform. . . . Focusing attention on managing for environmental results could assure that resources are aimed at real environmental improvements and could create incentives for management improvement.
>
> —Al Alm, Former Deputy Administrator, EPA

Management is an evasive term. In general, it can be considered a toolbox of methods and techniques used to realize productivity enhancement, cost minimization, and orderly transition from one time period to another.

When used in reference to hazardous materials, the term "management" can mean the selection of the best treatment alternative for a particular substance, or it can denote the policymaking philosophy of a corporate board of directors. This chapter will address management as a function of program administration. The technical alternatives including storage, treatment, and disposal will be discussed under the topic of substance management in Chapter 10.

GOALS

Proper administration of a hazardous materials management program should include the definition of specific goals. Four goals that should be common to all hazardous materials management plans will be discussed: regulatory compliance, preserving environmental quality, safety, and cost effectiveness. These goals are interrelated, and their prioritization will differ on a case-by-case basis.

Regulatory compliance is discussed in detail in Chapter 12. Ideally, compliance would ensure that no significant adverse environmental effects occur, but this is not always the case. Compliance with regulatory standards will not necessarily shield a company from certain types of liability if environmental degradation occurs. Thus, the manager of a hazardous materials program must often address issues beyond fundamental regulatory compliance.

It is desirable to avoid or mitigate any significant adverse effects to the environment. Motivation to preserve environmental quality extends beyond altruistic desires, as degradation of the environment can have direct results on a facility's public image and thus on its profitability. Any industry planning an expansion should be aware that its environmental track record will be subject to intense scrutiny. Poor public relations can result in permitting delays, increased costs, and residual community resentment. Abandoned waste sites and hazardous material discharges have social, political, and psychological impacts that cannot be ignored and that eventually translate into detrimental economic impacts.

Many studies have demonstrated the cost effectiveness of safety programs. The goal of a safety program is to decrease all risks to an acceptable level. Because the significance of any risk is a function of both severity and frequency, a comprehensive safety program aims to reduce both. Traditional approaches to chemical safety include the following:

1. Prevention of accidental releases of hazardous materials;
2. Prevention of human exposure in the event of a release; and
3. Prevention or minimization of injury in the event of exposure.

Steps 1 and 2 serve to minimize the probability of the occurrence of a harmful event, and step 3 addresses minimizing the severity of the potential effects.

A concern with employee safety is usually concurrent with a positive attitude toward regulatory compliance. The safety aspect of staff performance is an important factor to be considered in any efforts to maximize future business success and minimize regulatory compliance problems.

As with preserving environmental quality, compliance with regulations may provide a foundation for achieving desirable safety practices but may not be sufficient in and of itself. Safety programs beyond regulatory requirements are good management practices and often provide a more productive and secure workplace environment. A 1987 federal appeals court determined that employers have a general responsibility to protect workers against known hazards, whether or not specific regulatory standards exist.

However, no safety program can reduce all risks to zero. In realistic terms, at least some of the definition and scope of any safety program will be based on overall profitability. From a business point of view, the acceptable level of risk at any specific facility may become defined as the point where the cost of further decreasing a given risk is greater than the predicted cost assigned to the harm resulting from that risk. Such judgments are necessarily based on economics.

A good program will be managed cost effectively. This includes such obvious concerns as avoiding fines for noncompliance, selecting appropriate substance management technologies, and effectively utilizing personnel resources. Also to be considered are arrangements for adequate reserve accounts and insurance coverage.

Financial management decisions are usually based on some type of cost-benefit analysis. Such analyses must often address complex questions when investigating hazardous materials management strategies:

1. How does individual loss of life or impairment of function weigh against the cost to the population as a whole?
2. How can each potentially affected "statistical life" be assigned a dollar value?
3. Given limited funds, how does prioritization of the control of one substance or set of circumstances trade off with the control of others?
4. Is the public willing to accept certain risks in return for potential technological benefits?
5. Can the risks be quantitatively worked into a cost-benefit analysis?
6. Can the present value of risks be properly accounted for when they involve decades and intergenerational periods?

A thorough investigation of such questions is often necessary in order to formulate an appropriate set of priorities of long-term financial planning.

The recent insurance crisis has introduced a whole new subset of financial analysis questions, as the waste management industry has witnessed the virtual disappearance of a competitive market to provide environmental insurance. To what extent must the public, business, and insurance companies assume financial responsibility for risks inherent in providing the high standard of living in an advanced society? Should these risks be shared and, if so, how? What role should the civil justice system play in determining risk sharing? The answers to these questions will ultimately affect the management of any hazardous materials program that involves insurance coverage and will directly affect the cost effectiveness of any such program.

Some of the answers to cost-benefit-related questions are addressed through risk assessment and risk management, where exposure data is extrapolated to yield predicted mortality rates and the health risks are then compared with the calculated economic cost of alleviating the health dangers. In an ideal world, it would be possible to generate precise economic projections. Perhaps the most accurate statement about any hazardous materials management cost-benefit analysis is that all have inherent deficiencies, but each may provide reasonable ranges of economic conclusions. Risk analysis and management are discussed further in the next section.

IMPLEMENTATION

The function of an effective hazardous materials management program is to identify and reduce major risks. This is accomplished by a two-step process: risk assessment followed by risk management. The final step of implementation of most hazardous materials management programs is communication of the conclusions. In the case of a permit application hearing or facility siting study, a complete plan of interactive disclosure constituting a public relations program is highly desirable.

Risk assessment takes into account the available information about exposure to hazardous materials and assesses further data needs. For example, an industrial risk assessment of a potential chemical exposure might address the following:

1. An estimate of the severity of the problem in relation to human health and the environment;
2. The likelihood of future regulation;
3. The likelihood of potential suits resulting from release into the environment; and
4. An estimate of corporate financial exposure.

The process of such a risk assessment will generally follow these steps:

1. Definition of the potential sources of contamination, concurrent with characterization of the contamination. This typically involves sampling followed by laboratory analyses.
2. Identification of the possible pathways for contaminants to move from their point of origin and the potential rates of movement.
3. Identification of potential receptors of exposure to the contamination, including human, animal, and plant populations.
4. Determination of potential health impact to exposed populations.
5. Determination of acceptable levels of exposure.
6. Identification and evaluation of technologies suitable for reducing existing contamination or adequately minimizing or controlling potential releases.

While more data can always be gathered on any given subject, the inadequacy of data on hazardous materials management may be of such magnitude that it obscures the scope, nature, and complexity of the problems to be solved. Little is known about the long-term effects of hazardous materials on human health or the environment; thus quantitative risk assessment depends to a great deal on the utilization of safety factors and best engineering judgment. More than one disparaging analysis of risk assessment has concluded that it is equally subject to both statistical aberration and political manipulation.

However, risk assessment remains the only logical, rational, and methodologically consistent approach to making decisions in the highly sensitive area of hazardous materials management. And, as Lord Rothschild once said, "There is no point in getting into a panic about the risks of life until you have compared the risks which worry you with those that don't but perhaps should."

The risk management process involves selecting a course of action based on the risk assessment. This can range from taking no further action to implementing a plan to reduce the risk to a predefined level, which may include the following alternatives: regulatory compliance actions; waste minimization; recycling; training programs; legal actions; clean-up actions; insurance coverage; and changes in management practices, in business operations, substance management technologies, construction practices, and process operations. A proactive management program, including release detection systems, contigency planning, and monitoring, can reduce the potential costs of hazardous substance releases and can substantially reduce the cost of insurance or other required financial assurances. Note that liability may be reduced every time corporate management uses expertise in planning and making decisions but is never completely eliminated.

Of all the potential components of a course of action, monitoring is the one most likely to be a part of all risk management plans. Monitoring is the single most effective and reliable way to measure the level of protection attained by a given management program, and monitoring is the most widely applicable method of judging the effectiveness of selected action alternatives.

The overall effectiveness of a hazardous materials management program can be enhanced by clear communication with the people analyzing project impact, whether these be in-house technical staff, management personnel, agency staff, or the general public. In situations involving public input, a planned approach to public relations is very desirable. The goal of such a plan is to turn extemporaneous response into thoughtful, constructive input.

Poor public relations can be the cause of delays, increased costs, and residual community resentment. A technically and economically sound management plan may never be implemented unless it has the support of the involved community. A rule of thumb is that the more visible and significant the project, the more active the public relations program needs to be.

The general rules of any public relations program apply:

1. Provide a comprehensive source of information so that citizens and public and special interest groups will have convenient access;
2. Provide public education on the issues so that the public will be able to evaluate and comment on the proposed action;
3. Brief the local press regularly;
4. Establish two-way communications through open meetings and workshops; and

5. Where possible, establish one or two individuals as the central contact for information for the sake of continuity and personalization.

It is advisable to rely heavily on the results of the risk assessment and to make certain that the procedures and findings are conveyed in easily understandable terms. Explanation of the risk assessment process lends a degree of thoroughness and credibility to hazard-related analysis, which is crucial if the general public is to accept the recommended course of action.

Many firms embark on a program of information disclosure as a necessary evil, participating only because of regulatory requirements for public hearings or because of media pressure. But implementing a well-planned public relations program can build confidence in a facility's perceived environmental attitude, and sometimes actions set in motion by public participation make good economic sense. Reorganization of manufacturing processes, substitution of less hazardous raw materials, and recycling, recovery, or minimization of waste production can reduce operating costs considerably. When such changes occur without public input, they are certainly worthwhile to publicize.

INNOVATIVE APPROACHES

There are almost as many approaches to management strategies as there are definitions of the term "management." For this discussion, five current innovative practices which are pertinent to hazardous materials management have been selected: environmental audits; management inventories; computerized information management; coalition efforts by industries, agencies, environmental groups, and the public; and selected potential solutions to the insurance crisis.

The main goal of an environmental audit is to ensure regulatory compliance, although an audit may address other management goals discussed previously. An environmental audit is a formal self-appraisal conducted to assess a facility's compliance with environmental regulations and to reduce environmental problems and liabilities.

For the purposes of program management, an environmental audit can be expanded to address management policies, attitudes, and guidelines. A thorough environmental audit can provide guidance for preventing environmental hazards not currently addressed by regulation. It can also provide environmental guidelines concerning the definition of training programs, communication policy of facility personnel, early detection of developing problems, procedures for alerting personnel to emerging environmental developments (both technical and regulatory), and the implementation of high environmental standards with outside contractors. Although an environmental audit can be a valuable tool in program management, as stated previously, its main goal is to ensure regulatory compliance. Therefore, audits are discussed further in Chapter 12.

The process of taking an inventory is usually thought to be applicable to materials and supplies, and, in fact, such a physical inventory is often a component of an environmental audit. However, a management inventory can summarize management resources using many of the common inventory indices: operational effectiveness, current applicability, downtime, and replacement cost. Whereas an environmental audit deals with facility-specific information, a management inventory addresses management strategies and the skills of individual supervisors. It identifies responsibilities by position and activity. A well-conducted management inventory can identify redundancies as well as deficiencies because it serves to differentiate the role from the person filling it.

Computerized information management systems can aid in the storing and analyzing of the collected data. All the major environmental regulations require facilities to collect, analyze, and report on a wide range of data, and the majority of this information is chemical in nature. Environmental decision making requires the data to be arranged suitably for human pattern analyses and often requires numerous rearrangements of this data to evaluate various scenarios.

The foundation of such an information management system is an environmental data base, which is a permanent collection of both descriptive and routinely monitored data. The data base consists of files, each storing related data on a particular subject known as a data field. Once the desired data fields have been identified and the data comprising the files have been entered, users have the capability to manipulate and update enormous amounts of data rapidly.

In addition to information storage and management, pertinent software is available and is continuously being developed. Such programs can perform various functions, ranging from updating compliance requirements to simulating environmental incidents such as spills, permit violations, and inspections. A selected list of currently available software programs is shown in Table 9.1. The value of any computer-aided management program will, of course, depend a great deal on the identification and selection of an appropriate system, followed by careful implementation of the system.

Clean Sites, Inc., is a private group that uses an innovative approach to hazardous materials management. It is a coalition of industry and environmental groups which does not get involved in the actual cleanup of sites, nor does it participate in legal proceedings. Instead, it negotiates settlement between involved parties as an independent third party. The approach taken by Clean Sites has proven so effective that EPA has requested its aid in solving disputes at some of the country's worst dumps. In light of the previous discussion about computer systems, it is interesting to note that part of Clean Sites' success has been attributed to its work in developing sophisticated computer data bases to aid in determining who is responsible for what share of clean-up costs.

Table 9.1
Selected Hazardous Materials Management Programs

Program Name	Addresses	Functions
Anasoft	Environmental data	Monitors flows, PH, pollution levels etc.
Hazard	Uniform hazardous wastes manifests	Prepares manifests, tracks wastes
PCB Hazard	PCB containing transformers	Assesses hazard, outlines options, provides cost analysis
Q-Tracker	Environmental data, chemical data, regulatory requirements	Tracks permit renewals, violation reports, generates various incident reports
Safechem	Hazardous chemicals	Provides extensive information on more than 900 chemicals
Sludge Manager	Sludge and by-product utilization	Database management and record-keeping
Spilcom	Contingency planning	Aids notification procedure, facilitates insurance planning
Toxic Alert	Regulatory requirements, esp. "Right-to-Know"	Generates MSDS sheets
Waste Documentation and Control	Waste disposal and transportation	Generates required forms, maintains related data lists

Clean Sites shows by example that a beneficial hazardous materials management strategy often consists of nonconfrontational involvement of all affected parties. Smaller facilities practice the same principles that are used by Clean Sites when they engage in negotiation meetings with regulatory agencies and when they solicit the participation of involved community members in mitigating environmental impacts.

Another innovative approach involving a coalition effort between government and the public was attempted by Michigan in 1986. The state placed newspaper advertisements urging anyone with information about contamination at a specific site to contact the State Department of Natural Resources. The effort resulted in letters containing new and valuable information which aided clean-up efforts.

The current limited availability and high premiums associated with pollution liability insurance has provided vast incentives to seek solutions to the insurance crisis. The severity of the insurance crises is indisputable. In addition to federal liability insurance regulation, some states require levels of insurance coverage which are significantly higher than those required by EPA.

Numerous states are proposing or enacting legislation that will increase regulation of the insurance industry and is intended to ease the crisis by making more coverage available at lower premiums. Many states are exploring the use of state-funded insurance pools, and suggestions have been made that if programs to clean up toxic waste and asbestos are to succeed, the federal government may have to become the insurer of last resort.

In the private sector, professional groups and even individual installations are exploring the possibilities of pooling or sharing of risks or of individual self-insurance. The formation of such risk retention groups requires substantial financial investments, definition of group eligibility criteria, and the establishment of required administrative procedures. While it is not an answer to the underlying causes that have brought about the insurance crisis, self-insurance or risk retention groups provide a viable alternative while long-term reforms are being sought.

FUTURE NEEDS

The interrelation of the management tools discussed in this chapter is clearly evident. If a computerized information system can aid a facility in establishing a spotless environmental track record, it is likely that the facility can obtain lower insurance premiums. This, in turn, can leave more operating funds available to devote to a hazardous materials management program.

The foundation of hazardous materials management programs is the risk assessment process, and developments which improve this process will benefit all aspects of hazardous materials management. High on the "wish list" are the following:

1. Development of improved analytical tests for both on-site and laboratory use;
2. More substantial data base on the toxicological effects of chemicals on humans and the environment;
3. Increased research and development of technologies potentially suitable to control pollution sources; and
4. More public education about the perspective of involved risks compared with daily societal risks.

Risk assessment is a growing science that is here to stay, and one that requires the cooperative efforts of many specialists. It is hoped that such interdisciplinary team efforts will yield significant advances in the process.

Defining uncertainties isn't a substitute for finding ways of dealing with them. But risk-based management focuses on the broad range of uncertainties that bear on selecting a course of action and can help achieve a desirable balance among economic benefits, engineering features, environmental values, and long-term environmental impacts.

10

SUBSTANCE MANAGEMENT

Dilution is not a solution.

—Old engineering graduate school adage

The need to properly manage hazardous materials and dispose of hazardous wastes has been with us for more than 100 years. Within the last fifteen years or so, this need has become more evident. Much of this is due to the realization of the effects of past management practices, many of which were adopted out of hand or in compliance with regulations existing at the time, with no evaluation of the eventual results.

In the past few decades, the amount of waste produced per pound of product manufactured has significantly decreased for most processes. However, the enormous increase in national production has also significantly increased the amount of waste to be managed.

Regardless of the intent of environmental policies or the goals of environmental legislation, the management of hazardous substances is a problem that is inherent in our lifestyle. Hazardous materials are used in the production of everyday products and are the by-products of goods, clothes, foods, cars, and so forth—some of which are listed in Table 10.1. Given the ubiquitous nature of the problem, the next step in achieving a solution is the selection of appropriate management options.

Specific hazardous materials management technologies should be selected based on the hazards involved, available technology, regulatory requirements, cost efficiency, and numerous other factors. The alternative technologies can be grouped into the following categories:

Storage

Minimization/Reduction

Recycling/Reuse/Recovery/Exchange
Delisting
Treatment
Disposal

All are applicable to hazardous waste, but only storage, minimization, and recycling are generally applicable to materials. Some of the management options reduce the volume and/or the hazardous characteristics of the material in question. Others do not. These relationships are depicted in Figure 10.1. It is the technologies that do not actually change or destroy the substance, such as disposal, that are usually of the most public concern, often justifiably.

Table 10.1
Hazardous Wastes from Everyday Products

Products People Need and Use	The Potentially Hazardous Waste They Generate
Plastics	Organic chlorine compounds
Pesticides	Organic chlorine and organic phosphate compounds
Medicines	Organic solvents and residues, and heavy metals (e.g. mercury and zinc)
Paints	Pigments, solvents, heavy metals, and organic residues
Oil, gasoline, and petroleum products	Oil, phenols, and other organic compounds; heavy metals; ammonia salts; and caustics
Metals	Heavy metals, fluorides, cyanides, acid and alkaline cleaners, phenols, solvents, pigments, oils, abrasives, and plating salts
Leather	Heavy metals and organic solvents
Textiles	Heavy metals, dyes, organic chlorine compounds, and solvents

Figure 10.1
Hazardous Materials Management Technologies

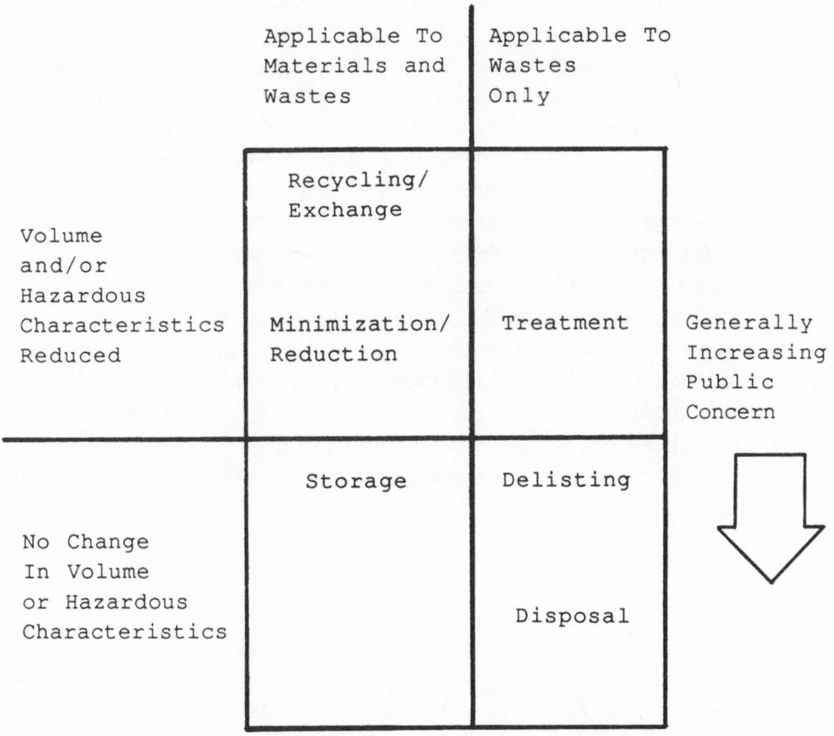

	Applicable To Materials and Wastes	Applicable To Wastes Only	
Volume and/or Hazardous Characteristics Reduced	Recycling/ Exchange Minimization/ Reduction	Treatment	Generally Increasing Public Concern
No Change In Volume or Hazardous Characteristics	Storage	Delisting Disposal	

STORAGE

Storage is the holding of hazardous materials prior to use, transport, recycling treatment, or disposal. In the case of waste materials, retrievable storage may provide an alternative for wastes which have been determined to be unsuitable for landfills, but for which no current economical treatment technology exists. To date, such a situation has necessitated the long-term storage of many PCB wastes.

Retrievable storage systems can include tanks, drums, lined holding ponds, and other conventional or specialized containers. Storage of radioactive materials or wastes may involve lead-filled canisters and may necessitate remote control handling apparatus.

Although the general meaning of the term "storage" seems obvious, an exact definition of regulatory terms is necessary to determine which facilities accumulating wastes are subject to storage facility regulations. According to the Resource Conservation and Recovery Act (RCRA), a generator may accumulate hazardous wastes for a period of up to 90 days before the facility is classified as a storage facility. Small quantity generators are allowed a "grace period" of 180 days, which may be extended to 270 days for certain isolated facilities. Note that this definition applies only to RCRA-regulated wastes. The storage of hazardous materials is generally not as stringently regulated as the storage of hazardous wastes.

Concerns about storage facilities include leakage of wastes into ground and surface water, accidental leakage from containers into the environment, and fires and explosions that may occur in facilities storing ignitable or reactive wastes. Many of these concerns can be addressed by following the proper procedures for storage of materials, such as not storing incompatible chemicals in dangerous proximity. Table 10.2 lists fundamental recommended storage and handling procedures. Table A.16 in the appendix is an extensive list of incompatible chemicals, which should not be stored together unless special precautions are taken.

A major issue requiring evaluation prior to designating a storage facility is whether immediate use, transport, treatment, or disposal would minimize the risk resulting from the transfer of the material to and from storage. A process analysis initiated by this question can often eliminate or reduce the need for storage facilities.

Table 10.2
Proper Procedures for Storage of Hazardous Materials

- Use personal protective equipment
- Be familiar with specific hazards of material being handled
- Obey all safety rules
- Do not smoke while handling materials
- Store chemicals according to manufacturer´s instructions, away from other chemicals or environmental conditions which could cause reactions
- Face labels on containers out
- Keep stacks straight and aligned
- Check for location accuracy
- Do not stack containers too high
- Check for loose closures
- Place into proper location as soon as possible
- Do not block exits or emergency equipment
- Report all spills or leaks immediately

MINIMIZATION/REDUCTION

Reduction of hazardous waste is actively promoted by the EPA as a preferred substance management option, and the 1984 amendments to RCRA require minimization to be addressed on manifest forms, program descriptions, and progress reports. A 1986 study by the Office of Technology Assessment (OTA) of the U.S. Congress concluded that, as a means of pollution prevention, reducing the generation of hazardous waste at its source is the best and most effective way of reducing risks to health and the environment. In a 1987 followup, EPA issued a five-part strategy intended to remove technical barriers to reducing hazardous waste.

Methods of hazardous waste reduction include process modifications, raw material substitutions, source segregation, and efficient prevention or management of spills and leaks. Process modifications and raw material substitutions can achieve reduced use of hazardous materials as well as reduced generation of hazardous wastes. These two practices are often interrelated and are usually implemented to achieve long-term utilization of relatively nonhazardous material where a hazardous material has historically been used. Source segregation prevents generation of large volumes of hazardous waste by preventing contamination of one waste by contact with another.

Restraints imposed by product quality requirements may prevent the successful implementation of process modifications or material substitutions. Even where minimization/reduction actions are applicable, these technologies have historically been resisted by industry, because their implementation is generally costly and must be tailored individually to the characteristics of each specific facility. Implementation often requires the substitution of process equipment designed to provide for chemical reuse rather than once-through processing. However, the cost-benefit analyses of such practices is changing because of increased regulation, stiffer penalties for noncompliance, and increased costs of hazardous waste treatment or disposal.

Ideally, waste minimization/reduction can benefit industry, the public, and the regulatory agencies by reducing production costs and related liabilities, providing better environmental protection, and reducing the need for enforcement activities.

RECYCLING/REUSE/RECOVERY/EXCHANGE

Recycling hazardous materials can result in a reduction or total elimination of the generation of hazardous waste. Related terms are recovery and reuse.

An exchange is essentially a reuse function where more than one facility is involved. An exchange matches one industry's output to the input requirements of another. Waste exchange organizations act as brokers of hazardous materials by purchasing hazardous materials and transporting them as a resource to another client. Waste exchanges commonly deal in solvents, oils, concentrated acids and alkalis, and catalysts. Limitations affecting the feasibility of such exchanges include transport distance, purity of the exchange product, and the reliability of supply and demand.

DELISTING

Industry can petition EPA to "delist" a listed waste, based upon the presentation of evidence that the waste is not hazardous. This process does not affect the quantity or characteristics of the waste in question, but delisting eliminates many of the regulatory requirements pertaining to the waste. This, in turn, can significantly simplify treatment or disposal requirements.

The delisting process is not a simple or quick procedure; it requires substantial documentation. The 1984 amendments to RCRA require consideration of factors in addition to those constituents for which the waste was originally listed. In 1985, EPA stated its intention to evaluate a waste with regard to any other hazardous constituents that may reasonably be expected to be present. These criteria are to be applied to wastes previously delisted in addition to new delisting petitions. These new procedures will probably significantly reduce the number of wastes that are successfully delisted.

TREATMENT

In order to achieve regulatory compliance for materials which cannot be managed by the previously discussed options, treatment or disposal technologies must be implemented. In a broad sense, hazardous waste treatment can encompass both the operation of substance management facilities and proper disposal. For the purposes of this discussion, a narrower, more technical definition of treatment is used: any process that changes the physical or chemical nature of a waste in order to render it less hazardous, reduce its volume, or make it easier to handle. The various treatment technologies fall into four major categories: physical treatment, chemical treatment, biological treatment, and incineration.

Physical Treatment

Physical treatment consists of nonchemical methods designed to reduce the volume of waste or to separate out a liquid portion of the waste. It does not result in the destruction of the toxic components of the waste. Some physical processes may be used to reclaim or recover valuable, reusable components of the material before they are discarded. Examples of physical treatment include solid-liquid separation processes, membrane separation processes, evaporation, distillation, stripping, solvent extraction, absorption, and adsorption.

Solid-liquid separation processes involve the physical segregation of the solid and liquid components of a waste stream. These processes include screening, sedimentation, flotation, filtration, and centrifugation.

Screening is widely used to remove large particles from waste water. Sedimentation is the removal of solid particles suspended in a liquid by gravity settling. When solids are so finely dispersed that they do not settle easily, chemical coagulating or flocculating agents may be added to help agglomerate the particles together to speed settling. Since sedimentation is often performed in large open tanks or holding ponds, there may be significant air emissions when high concentrations of volatile components are present in the waste stream.

Dissolved air flotation separates solids from liquids by the attachment of tiny air bubbles to the solid particles. Filtration involves passing a mixture of solids and liquids through a porous medium which intercepts the solids. In centrifugation, the solid and liquid components of a waste stream are separated by rapidly rotating in a vessel.

Membrane separation processes are used to separate components of liquid solutions. They are based on the ability of membranes to allow the passage of one component in a liquid stream while effectively preventing passage of others. Membrane separation processes include dialysis, reverse osmosis, ultrafiltration, and electrodialysis.

Dialysis is a membrane separation technology that uses osmosis as the primary driving force for transport of components from one side of the membrane to the other. Reverse osmosis is similar to dialysis, but mechanical force (high-pressure pumping) is the primary driving force for separation. The membrane in reverse osmosis allows passage of the solvent (usually water) but not the dissolved components of the solution. Reverse osmosis can remove metals and some organics from a waste stream, depending on the molecular weight and size of the particles to be separated.

Ultrafiltration, which is similar to reverse osmosis, allows low molecular weight components such as salts to pass through the membrane with the water phase. Electrodialysis is a membrane separation process that relies on electric forces as the primary driving force for transport and uses membranes which selectively allow the passage of ionic components of an aqueous brine stream. This results in a concentration of the ionic components and a purification of the water.

Evaporation is a physical process involving the vaporization of a liquid from a solution or a slurry. The usual objective of evaporation is to concentrate the solution or slurry by driving off some of the volatile liquid. The vaporized liquid may or may not be recovered. In distillation, boiling is used to separate liquid components or trapped solids from a liquid stream. Distillation is frequently used for recycling used solvents.

Air or steam stripping is a technique for separating components of a liquid mixture. In air stripping, components with high vapor pressure and low solubility are volatized during contact with air. This process is especailly applicable to certain organics. Steam stripping accomplishes separation by the introduction of heat in the form of steam. This process is generally used only when the organic components and water are incapable of mixing.

Solvent extraction is a process for removing organic substances from aqueous streams. In the process, a solvent that is immiscible with the liquid waste stream is mixed with the stream. When this solvent is removed from the mixture, it extracts organics from the original stream.

In absorption, spongelike material, such as sawdust, is used to soak up free liquid. Absorption is frequently used when there is free-standing liquid in containerized wastes, or when spills of hazardous liquids occur.

Adsorption differs from absorption in that it is a process of surface retention of one material by another, as opposed to retention accomplished by penetration of one material into the bulk of another. A common application of adsorption is the use of activated carbon to remove organics from a waste stream.

Chemical Treatment

Some of the more common forms of chemical treatment include neutralization, precipitation, oxidation/reduction, ion exchange, chemical dechlorination, and solidification.

Neutralization converts acid or alkaline wastes to substances which are no longer highly corrosive. Neutralization is one process where combining two hazardous waste streams can potentially yield a nonhazardous material. The products of neutralization are usually water and salt.

Precipitation is a chemical process for removing dissolved components from a solution by altering the equilibrium relationships affecting the solubility of the components. The alteration can be achieved by changing the pH, chemical reactions which convert the components to insoluble products, and/or by changing the temperature of a saturated or near-saturated solution to decrease solubility of the components. Precipitation is often used to accomplish the removal of heavy metals and is usually used in combination with some type of solid removal process such as sedimentation, flotation, centrifugation, or filtration.

Chemical oxidation/reduction processes use a chemical reaction to detoxify hazardous wastes. Electrons are transferred from one reactant to the other, and in the process chemical bonds may be broken, thus converting a toxic material into simpler, less toxic chemicals. Some commonly used oxidizing agents are ozone, air, chlorine gas, chlorine dioxide, sodium hypochlorite, calcium hypochlorite, potassium permanganate, hydrogen peroxide, and nitrous acid. Other easily oxidized compounds such as sulfites, sulfides, and simple organics can be oxidized simply by passing air or oxygen through the solution.

Ion exchange is a process that separates dissolved inorganic material from an aqueous liquid. The liquid is passed through a fixed bed containing a natural or synthetic resin. The resin exchanges ions with the inorganics in the solution, resulting in the inorganics being removed from the solution and attaching to the resin column. Ion exchange is recommended for acid solutions containing certain metals, salt solutions, and heavy metals in aqueous solutions. Electroplating waste streams containing chromium and cyanide wastes, and mixed waste streams from metal finishing operations are typically amenable to treatment by ion exchange. Ion exchange can also be used to remove radioactive impurities from certain materials.

Chemical dechlorination refers to a group of emerging technologies which can be used to strip chlorine atoms from highly chlorinated toxic compounds, such as PCBs, in order to produce a nontoxic residue. The major concerns detracting from this technology to date have been economic, rather than environmental. The cost of treating the resulting process reagents containing high concentrations of PCBs has been prohibitive. Broadening the applicability of the process to treat other highly chlorinated compounds such as pesticides may help make the technology more economical.

Solidification, also known as stabilization, can be a purely physical process in some cases, but usually involves the addition of chemicals. The resulting cementlike chemical reactions convert liquid or semi-liquid wastes into a solid mass. This process is intended to immobilize hazardous constituents by retaining them in the material in their most insoluble form. This, in turn, prevents the pollutants from migrating further into the environment.

Stabilization/solidification processes can be categorized as follows:

— Cement-based Process: The wastes are stirred in water and mixed directly with cement. The suspended particles are incorporated into the hardened concrete.

— Pozzolanic Process: The wastes are mixed with fine grained silicicous (pozzolanic) material and water to produce a concretelike solid. The most commonly used materials are fly ash, ground blast furnace slag, and cement kiln dust.

— Thermoplastic Techniques: The waste is dried, heated, and dispersed through a heated plastic structure. The mixture is then cooled to solidify the mass.

— Organic Polymer Techniques: The wastes are mixed with a prepolymer in a batch process with a catalyst. Mixing is terminated before a polymer is formed, and the spongy resin-mixture is transferred to a waste receptacle. Solid particles are trapped in this spongy mass.

— Surface Encapsulation: The wastes are pressed or bonded together and enclosed in a coating or jacket of inert material.

Table 10.3 summarizes some basic data on the applicability of various stabilization processes. Similar immobilization techniques are used for a variety of radioactive wastes.

Table 10.3
Stabilization/Solidification Processes

Fixation System	Major Materials Stabilized	Materials to Which System is not Applicable
Cement-based	Toxic inorganic wastes, stack gas, scrubbing waste	Organic wastes, toxic anions
Pozzolanic	Toxic inorganic industrial wastes, stack gas, scrubbing wastes	Organic wastes, toxic anions
Thermo-plastic	Toxic inorganic industrial wastes	Organic wastes, strong oxidizers
Organic polymer	Toxic inorganic industrial wastes	Acidic materials, organics and strong oxidizers
Encapsulation	Toxic and soluble inorganic industrial wastes	Strong oxidizers

Biological Treatment

Biological treatment utilizes microorganisms to consume wastes. Under certain properly controlled conditions, biological treatment can convert hazardous constituents in liquid waste streams, soils, or sludges into nonhazardous sludges, gas, and water. Many of these processes are similar to those used in municipal wastewater treatment systems, such as trickling filters, activated sludge, aerated lagoons, waste stabilization ponds, and anaerobic digestors. Land farming and the development of special microbial strains are two additional types of biological treatment processes that are applicable to certain hazardous wastes.

In the trickling filter process, wastes are allowed to trickle through a bed of rock or synthetic media coated with a slime of microorganic growth. Microorganisms decompose the organic matter in the waste stream. Open tanks or towers house the filter, upon which the microorganisms grow, and open tanks clarify the filter effluent. The filter tank or tower uses a rotating spray system for feeding the waste stream to the filter surface.

The activated sludge process decomposes organic wastes in waste water by exposure to biological growth. Water is continuously recycled in the system to maintain high concentrations of microorganisms.. The process involves an aeration step to provide the oxygen in the waste streams necessary for biological decomposition, followed by a clarifier in which the sludge is separated from the organic free water. A portion of the sludge from the clarifier is recycled back into the aeration tank to maintain high concentrations of microorganisms.

Aerated lagoons are based on the same biological decomposition principle as activated sludge systems. They consist of large earthen lagoons containing high concentrations of microorganisms and waste water, which is agitated to increase the oxygen content to encourage decomposition. The lagoon step is usually followed by a process where sludge settles out of the waste. Organic decomposition typically takes longer in aerated lagoons than in activated sludge systems, where the sludge is recycled.

Waste stabilization ponds are large shallow ponds in which trace levels of organics are decomposed over a long period of time. Aeration is provided only by wind action. In deeper ponds anaerobic digestion takes place. Stabilization ponds have been widely used to provide a final polishing of waste water to insure that effluent standards can be met.

In anaerobic digestors, microbes decompose organic matter in closed vessels in the absence of air. This process has traditionally been a supporting process to produce energy for other sludge-producing biological processes. An end product from anaerobic digestion is methane, a combustible gas.

Land farming is a waste treatment technique that uses microorganisms naturally occurring in the soil to biodegrade organic wastes. In a landfarming operation, the waste is either applied on top of land which has been disked or injected 4 to 6 inches beneath the surface. (Of course, in applications involving the reclamation of contaminated sites, the waste is already in place.) The land is plowed periodically to increase the oxygen content needed by the microorganisms to effectively biodegrade the wastes. The technique is also called land spreading, sludge farming, solids incorporation, and land treatment.

The development of specialized microbial strains has proven valuable where specialized wastes are being degraded. Special cultures of naturally occurring bacteria which have the capability to decompose or digest such materials as aromatic hydrocarbons, fats and greases, and ammonia have been developed. These mutant bacteria can degrade much more material than their naturally occurring precursors. Such specialized bacterial cultures have been applied to batch treatment processes where concentrated wastes have been isolated in spill ponds or equalization tanks, and with some success to wastes from chemical plants, refineries, petrochemical complexes, and textile plants. A special culture has also been developed that can remove cyanide toxins from chemical plant waste waters. These bacteria are presently being marketed by several companies within the United States.

Incineration

Incineration is the controlled burning of a waste at very high temperatures for a specified length of time to destroy combustible constituents in the waste. For some hazardous wastes, especially organics, incineration provides a preferred treatment alternative. For other hazardous wastes, such as certain heavy metals, incineration does not normally provide treatment. However, at least one new incineration technology is being developed which is intended to be applicable to heavy metals. Rather than resulting in combustion, the high temperatures theoretically cause metals in contaminated soils to bond to soil particulates, rendering the metals nonleachable.

Common incinerator types are listed in Table 10.4. Unlike most treatment technologies, the major contamination concern associated with incineration is with atmospheric, rather than aqueous, emissions. To determine the extent of atmospheric emissions, which have related health hazards, a hazardous waste incinerator must undergo a test burn. During the test, all operating conditions are closely monitored and documented. Emissions from the incinerator are analyzed for an indicator parameter, termed the Primary Organic Hazardous Constituent (POHC). A POHC is selected for a specific incinerator feed based on associated health hazards and the difficulty in achieving combustion. All other substances in the incinerator feed are then assumed to have been destroyed in a level equal to or greater than the level of POHC destruction. Despite extremely high destruction efficiencies, however, there are always some contaminants emitted to the air.

Other areas of concern pertaining to incineration include the following:

- Nonatmospheric Pollution: By-products, such as ash or scrubber residues, may be hazardous.
- Energy Requirements: Many mixed wastes are only marginally combustible, requiring a supplemental fuel. This results in an additional use of possibly scarce fuel.
- Aesthetics: Even if no significant pollution is occurring from incineration, there is usually a rather large steam plume which some find aesthetically unacceptable.

Table 10.4
Common Solid Waste Incinerator Types

Process	Description	Temp. Range ^oC	Application
Grate Type	Refractory or refractory lined box utilizing either a stationary or moving grate	Varies	Industrial trash, sludges
Rotary Kiln	Large cylinder rotating on trunnions	810-1650	Wide application
Multiple Hearth	Usually four or more horizontal refractory line hearths	315-870	Wide application
Fluidized Bed	Involves the use of a fluidizing material such as alumina which acts as a hearth	760-870	Petroleum and paper by-products, sludge
Liquid Injection	Vertical or horizontal units into which atomized waste is sprayed	650-1650	Liquid wastes, esp. chlorinated hydrocarbons

Pyrolysis is a treatment technique closely related to incineration, since both incineration and pyrolysis are processes for reducing the volume or toxicity of organic wastes by exposing them to high temperatures. In pyrolysis, also known as starved-air combustion, the waste is exposed to high temperatures without sufficient air to support combustion. As a result, the waste breaks down into other products which might be reclaimable or more cleanly combustible than the original waste. The equipment for this process is not unlike more typical incineration technology, but during the process, the air, which for most incineration processes is usually provided in excess, is limited. Incineration and pyrolysis are both used to treat certain radioactive wastes.

Summary

The increasing quantity of hazardous waste requiring treatment, along with the need to develop technologies capable of decontaminating abandoned sites, is the driving force behind technological advances developing new treatment processes and improving traditional ones. Table 10.5 lists some innovative treatment technologies, reflecting some of the current emphasis on in-situ technologies, PCB treatment, and sludge management. The long-term effects of new technologies and some variations of conventional technologies have yet to be determined.

Monitoring remains an important aspect of all treatment processes to ensure that the desired result is achieved. Additionally, all treatment process by-products or residual wastes require evaluation for appropriate treatment or disposal.

DISPOSAL

Disposal of hazardous wastes usually proves to be a more environmentally sensitive issue than either storage or treatment. The three main disposal alternatives are landfills, underground injection wells, and ocean dumping.

Landfills

Like all disposal technologies, landfilling does not assure permanent containment of hazardous wastes. Protecting groundwater from contamination by hazardous waste constituents is one of the most important considerations in land disposal. Some states have enacted laws prohibiting certain substances from being landfilled. EPA banned the landfilling of specific wastes through the enactment of the 1984 amendments to RCRA, and many more substances are expected to be barred from landfills in the future. These amendments call for an investigation of the entire concept of

Table 10.5
Selected Innovative Treatment/Clean-up Technologies

Category	Technology	Comments
General In-situ Cleanup	Bioreaclamation Soil vapor extraction Soil washing Air stripping Vitrification (soil melting) Activated carbon injection	Generally based on traditional treatment processes. Innovation involves application to soil and groundwater.
Metal Contaminated Soils Treatment	Bonding through material injection Incineration	Renders metals in soil non-leachable.
Molten Metal Destruction	Vaporization and detoxification	Applicable to nerve gas and selected pesticides. Potentially applicable to PCBs.
Sludge Management	Composting	Results in product recovery.
	Vapor compression	Results in dried product, potentially suitable for incineration.
	Downhole	Aqueous phase oxidation application, odor free.

landfilling hazardous wastes in order to determine the feasibility of prohibiting landfilling completely.

The concept of completely prohibiting landfilling is a complex one. Through the enactment of various regulations, especially the 1984 RCRA amendments, Congress has indicated a preference for treatment, destruction, and decontamination over short- or long-term land storage. Some believe it is a progressive measure to prohibit all landfilling. Others, however, believe the lack of economically accessible landfills directly contributes to the frequency of "midnight dumping," or the illegal disposal of waste. It will be difficult in practice to abolish landfilling until other economically feasible technologies for substance management are available.

In the meantime, significant advances have been made in landfill design and operation, many of which have been in response to regulatory requirements. To minimize and contain leachate generation, landfills can be constructed in discrete cells with impermeable liners and covers. Groundwater monitoring around each cell makes it possible to detect leaks in individual cells, which facilitates immediate response to potential pollution migration.

Underground Injection Wells

Underground injection involves the use of specially designed wells to inject wastes into deep earth strata containing non-potable water. Many injection wells are on the sites of oil production facilities. To prevent plugging of the injection equipment, wastes are usually pretreated to remove all solids greater than one micron in size. The depth of an injection well ranges from 1,000 to 8,000 feet and varies according to the geological factors of the area. The well must be constructed to assure that any potable water zones are isolated and protected. At a minimum this generally means that well casings must be cemented and must extend through all potable water zones.

Underground injection is a controversial disposal alternative because of the possibility of the contamination of usable groundwater above the injection zone. The following are issues of concern in underground injection of hazardous wastes: subsurface conditions, treatment of the wastes consistent with well design, verification of the mechanical integrity of wells, safeguards to ensure that the injected wastes do not migrate out of the wells, and groundwater monitoring.

Ocean Dumping

Ocean dumping, as defined in the Marine Protection, Research, and Sanctuaries Act (MPRSA), includes disposal of wastes in bulk or in containers, incineration at sea, and the burial of waste in the seabed.

Certain industrial and municipal wastes may be ocean dumped if they do not unreasonably degrade or endanger the marine environment or human health. The determination is made by EPA.

This disposal method has limited applicability to hazardous wastes, as it is generally not appropriate for wastes containing persistent toxic substances, particularly ones with a potential for bioaccumulation in the marine food chain. Significant amounts of low-level radioactive wastes were ocean dumped between 1946 and 1970. This practice has ended, mainly because of the perceived stringency of pertinent regulations. Currently, dumping low-level radioactive wastes requires an impact assessment and congressional approval prior to permit approval by EPA.

MANAGEMENT DURING CONSTRUCTION

Undersirable environmental consequences may occur at construction sites because protection during construction often receives less attention than the safeguards intended for an operating site. Regulatory agencies often hold the prime contractor accountable for environmental protection, and this liability can extend to problems at the site detected years later.

The same program management goals and substance management options discussed previously apply to construction sites. Common problems involve the lack of trained personnel and a knowledge of regulatory applicability. Simply because a project is under construction does not mean it is not under regulation; RCRA regulates chemical construction wastes transported off-site, and the CWA regulates most common construction discharges. Other federal and local regulations may also apply.

Hydrological and other geotechnical changes resulting from site preparation require evaluation. Land clearing and excavation can result in excessive runoff, and changes in the liquid pathways to and from wetlands can alter stormwater runoff. If contaminated with chemical wastes, the runoff can significantly degrade existing environmental conditions.

SUBSTANCE MANAGEMENT FACILITIES

Substance management facilities are a requirement of today's technological society, and although few communities desire to be the site of such facilities, they must be located somewhere. The emotional aspect of hazardous waste management has created an "anywhere but here" syndrome in many instances. This syndrome has become so widespread it even has its own acronym: NIMBY, which stands for "not in my backyard!"

The following excerpt, reprinted from *Engineering Times*, illustrates an extreme example of this syndrome.

Hazardous Waste Decision Provokes Talk of Secession

As federal and state officials have become painfully aware in recent years, nobody is especially receptive to the idea of having hazardous wastes disposed of near their homes. Now the rural western Massachusetts community of Brimfield has taken what some observers have referred to as the "Not-In-My-Backyard" syndrome to new extremes.

Resentful of the state's decision to site a hazardous waste treatment facility in their town, some Brimfield residents are suggesting that the community secede from the Commonwealth.

The protesters intend to have the secession question debated at Brimfield's town meeting next month.

According to a *New York Times* report, this is not the first time Brimfield, now with a population of about 2,400, has asserted its independence. In 1774, two years before the colonies had agreed to challenge British rule, Brimfield voted to declare its independence from Great Britain.

Obviously, siting management facilities is a complex issue which depends on a combination of environmental, social, and economic conditions. A handbook on *Siting Hazardous Waste Management Facilities* (National Audubon Society, the Conservation Foundation, and Chemical Manufacturers Association, 1983) recommends that any siting process

1. Be well thought out by the developers, facility operators, and government agencies;
2. Provide for early and full public information and participation;
3. Be understandable to the public;
4. Address the legitimate concerns of all parties;
5. Provide the public with accurate information concerning benefit, risks, and alternatives to a proposed facility;
6. Be a shared consultative process among site developers, facility operators, communities, government agencies, and other affected parties; and
7. Provide opportunities for all concerned to negotiate areas of dispute.

Furthermore, it is recommended that an evaluation of a proposed facility include an assessment of

1. The need for a facility of some sort;
2. Whether the proposed facility represents the most appropriate technology for handling the particular wastes; and
3. Whether the location proposed for such a facility is suitable given environmental, social, and economic concerns.

Improved substance management facilities, new substance management technologies, and improved processes for siting such facilities may do much to alleviate problems that have occurred in the past.

MANAGEMENT ON THE HOME FRONT

Dramatic technical advances, stricter regulations, and even the most exemplary materials management strategies by industry will not resolve environmental problems without active participation by the public. According to the Water Pollution Control Federation (WPCF), the average household contains between three and ten gallons of materials that are hazardous to human health or to the natural environment.

Like industries, the average household can implement practices to reduce the amounts of hazardous materials. The need for the product should be established before purchase, to aid in obtaining the right product and eliminate the need to purchase surplus amounts or certain types of chemicals. The consumer should become familiar with the directions on how to use the product and how to dispose of the container.

Table 10.6 gives guidelines concerning the disposal of potentially hazardous materials commonly found in the home. These materials should never be dumped on the ground, and sources of information such as Table 10.6 or regulatory agencies should be consulted before pouring such materials down the drain. Another source of information about proper product usage is provided at a toll-free number staffed by the Chemical Manufacturers Association (CMA): 1-800-CMA-8200. The service provides general information and referral to the manufacturers of specific chemical products.

Table 10.6
Household Hazardous Waste Chart

Probable Location of Generation	Type of Waste	Pour Down Drain With Plenty of Water	Sanitary Landfill	Hazardous Waste Contractor or Community Collection	Recycle
Kitchen	Aerosol cans (empty)	–	x	–	–
	Aluminum cleaners	x	–	–	–
	Ammonia based cleaners	x	–	–	–
	Bug sprays	–	–	x	x
	Drain cleaners	x	–	–	–
	Floor care products	–	–	x	–
	Furniture polish	–	–	x	–
	Metal polish	–	–	x	–
	Window cleaner	–	–	x	–
	Oven cleaner (lye base)	x	–	–	–
Bathroom	Alcohol based lotions (aftershaves, perfumes, etc.)	x	–	–	–
	Bathroom cleaners	x	–	–	–
	Depilatories	x	–	–	–
	Disinfectants	x	–	–	–
	Permanent lotions	x	–	–	–
	Hair relaxers	x	–	–	–
	Medicine (expired)	x	–	–	–
	Nail polish	–	x	–	–
	Nail polish remover	–	–	x	–
	Toilet bowl cleaner	x	–	–	–
	Tub and tile cleaners	x	–	–	–

Garage	Antifreeze	–	–	x	–
	Automatic transmission fluid	–	–	x	–
	Auto body repair products	–	x	–	–
	Battery acid (or battery)	–	–	x	–
	Brake fluid	–	–	x	–
	Car wax with solvent	–	–	x	–
	Diesel fuel	–	–	x	–
	Fuel oil	–	–	x	–
	Gasoline	–	–	x	–
	Kerosene	–	–	x	–
	Motor oil	–	–	x	–
	Other oils	–	–	x	–
	Windshield washer solution	x	–	–	–
Workshop	Paint brush cleaner with solvent	–	–	x	x
	Paint brush cleaner with TSP	x	–	–	–
	Aerosol cans (empty)	–	x	–	–
	Cutting oil	–	–	x	–
	Glue (solvent base)	–	–	x	–
	Glue (water base)	–	–	–	–
	Paint - latex	x	–	–	–
	Paint - oil base	–	–	x	–
	Paint - auto	–	–	x	–
	Paint - model	–	–	x	–
	Paint thinner	–	–	x	–

Table 10.6 (Continued)

Probable Location of Generation	Type of Waste	Pour Down Drain With Plenty of Water	Sanitary Landfill	Hazardous Waste Contractor or Community Collection	Recycle
Workshop (continued)	Paint stripper (lye base)	x	–	–	
	Primer	–	x	–	–
	Rust remover	–	–	x	–
	Turpentine	–	–	x	x
	Varnish	–	–	x	–
	Wood preservative	–	–	x	–
	Paint stripper	–	–	x	
Gardening	Fertilizer	–	–	x	–
	Fungicide	–	–	x	–
	Insecticide	–	–	x	–
	Rat poison	–	–	x	–
	Weed killer	–	–	x	–
Miscellaneous	Ammunition	–	–	x	–
	Artists paints, mediums	–	–	x	–
	Dry cleaning solvents	–	–	x	–
	Fiberglass epoxy	–	–	x	–
	Gun cleaning solvents	–	–	x	–
	Lighter fluid	–	–	x	–
	Mercury batteries	–	–	x	–
	Moth balls	–	–	x	–
	Old firearms	–	–	x	–
	Photographic chemicals (unmixed)	–	–	x	–
	Photographic chemicals (mixed and properly diluted)	x	–	–	–
	Shoe polish	–	x	–	–
	Swimming pool acid	–	–	x	–

11

EMERGENCY RESPONSE

> Whatever the emergency might be, trained individuals who work together can minimize the impact of emergencies on life, property, and economic stability . . . the capability to answer the call when a crisis occurs depends on how well they have planned with each other in mind.
>
> —Clyde A. Bragdon, Jr.
> Associate Director, Training and Fire Programs
> Federal Emergency Management Agency (FEMA)

Planning emergency procedures before an emergency occurs is known as preplanning or contingency planning. A good contingency plan is the backbone of proper emergency response.

Several practical considerations are applicable to all contingency plans. The plan must be specific, since generic plans seldom adequately address the unique conditions at a specific facility. The plan must address varied audiences, because, in addition to employees, others, such as regulatory agencies, fire officials, and emergency responders, may use the document. The language should be clear and concise, presenting duties, functions, decision conditions, and action alternatives in a straightforward manner. The document should be organized and formatted so that information is easy to find. Removable and tabbed sections are useful, as is segregation of pertinent regulations into an appendix. Last, but not least, the plan should be updated on a regular basis, especially whenever any major facility changes occur.

FEDERAL CONTINGENCY PLANNING

The basis of federal response to an emergency involving hazardous materials is the National Oil and Hazardous Substance Pollution Contingency Plan, which was first issued in 1968 as part of the requirements of the

Clean Water Act (CWA). The plan has been updated and revised numerous times, including revisions that were mandated by the Comprehensive Environmental Response, Compensation, and Liability Act (CERCLA).

The agencies and departments listed in Table 11.1 are among the entities that may be called upon to participate in an emergency incident involving hazardous materials. The two agencies designated as having primacy in emergency response situations involving nonradioactive hazardous materials are EPA and the U.S. Coast Guard (USCG). In the case of radiological accidents, federal responsibility is shared by the Federal Emergency Management Agency (FEMA), the Nuclear Regulatory Commission (NRC), and the Department of Energy (DOE). NRC and DOE maintain authority for planning and programs development for emergency response, notification, and technical assistance, advice, and involvement in response activities pertaining to radioactive materials incidents.

The National Contingency Plan (NCP) is implemented by coordinated efforts on the part of the National Response Team (NRT), the Regional Response Team (RRT), and the On-Scene Coordinator (OSC). The NRT is composed of representatives from the participating agencies and is located in the National Response Center in Washington, D.C. This center can be contacted 24 hours a day by calling the following toll-free number: 1-800-424-8802.

Table 11.1
Federal Agencies and Departments Governing Hazardous Materials

Acronym	Federal Agency
CEQ	Council on Environmental Quality
COE	U.S. Army Corps of Engineers
DHHS	Department of Health and Human Services
DOC	Department of Commerce
DOD	Department of Defense
DOE	Department of Energy
DOI	Department of the Interior
DOJ	Department of Justice
DOL	Department of Labor
DOS	Department of State
DOT	Department of Transportation
EPA	Environmental Protection Agency
FEMA	Federal Emergency Management Agency
FWS	U.S. Fish and Wildlife Service
MarAd	Maritime Administration
NMFS	National Marine Fisheries Service
NOAA	National Oceanic and Atmospheric Administration
NRC	Nuclear Regulatory Commission
USCG	U.S. Coast Guard
USDA	U.S. Department of Agriculture
USA	U.S. Army
USAF	U.S. Air Force
USN	U.S. Navy

RRTs, the regional counterparts of the NRT, are located in the ten regional EPA offices. The geographically appropriate RRT will be activated in the event of a major or potentially major discharge of hazardous materials. The OSC is the official who is chosen to direct and coordinate the activities at the site of the emergency incident. The interaction of the NRT, RRT, OSC, and response resources is depicted in Figure 11.1.

Figure 11.1
National Contingency Plan Operation

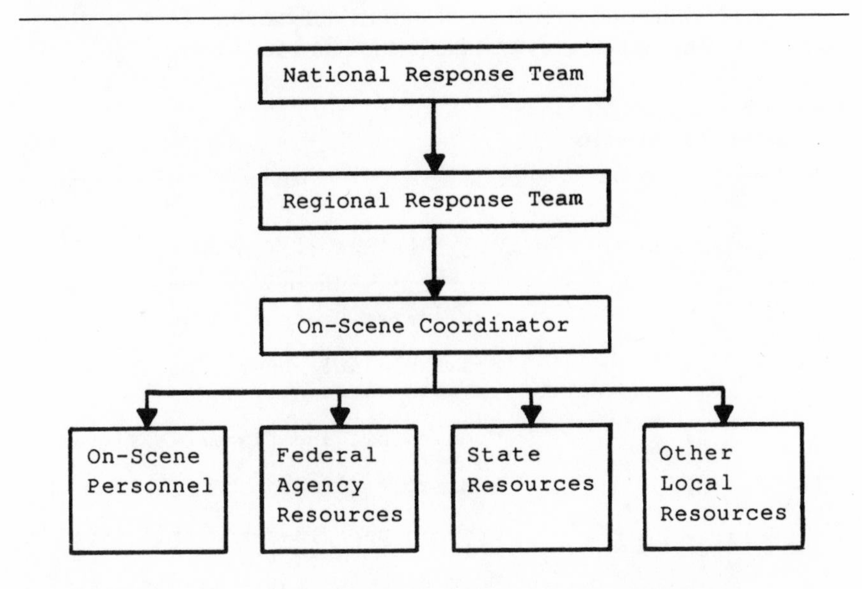

The actions of the response team members are dictated at the national level by the NCP, which in effect requires following the directions of appropriate, available site-specific contingency plans. The most useful document for handling a chemical emergency is often the contingency plan prepared for the specific facility where the emergency has occurred.

BASIC ELEMENTS OF A CONTINGENCY PLAN

A contingency plan may be a long, detailed document such as the NCP, or it may be a simple list of procedures applicable to one specific operation at one specific facility. Whether simple or complex, most contingency plans contain six basic elements. These are the procedures for identification, notification, evacuation, spill containment, fire protection, and documentation.

First priority is the identification of the hazardous material(s) involved and associated risks. The material in question must be identified before any action can or should be taken. Often, this identification is the major input into defining the extent of the emergency incident. Since the type of materials involved in a production emergency incident is usually apparent, the following discussion will center on the identification of materials in transit.

Readily accessible sources of identification should be investigated first. These include the Department of Transportation (DOT) labels/placards, shipping papers, and National Fire Protection Association (NFPA) 704 markings. The DOT system of labeling and placarding is discussed in Chapter 6. When applicable, placards will show a single-digit number in the lower corner which denotes the UN hazard designation. These numbers, and the DOT hazard classes associated with them, are shown in Table 11.2.

Table 11.2
UN Hazard Class Numbers

UN Hazard Class No.	Hazard Class
1	Explosives Class A, B, C
2	Flammable and non-flammable gases poison class A
3	Combustible and flammable liquids
4	Flammable solids
5	Organic peroxides and oxidizers
6	Poison B, irritating materials etiological materials
7	Radioactive materials
8	Corrosive materials
9	ORM class A, B, C, D, E materials

A summary of the NFPA 704 marking system is shown in Figure 11.2. Shaped like a diamond, the NFPA 704 symbol is divided into four smaller diamonds. The top three sections contain numbers ranging from 0 to 4 which describe the health hazards, flammability, and reactivity characteristics of the material. In all cases, a number 4 denotes the highest hazard. A cross-hatched letter "W" in the bottom diamond alerts fire-fighting personnel to the possible hazard in using water to extinguish a fire resulting from ignition of the material.

Figure 11.2
NFPA 704 Marking System

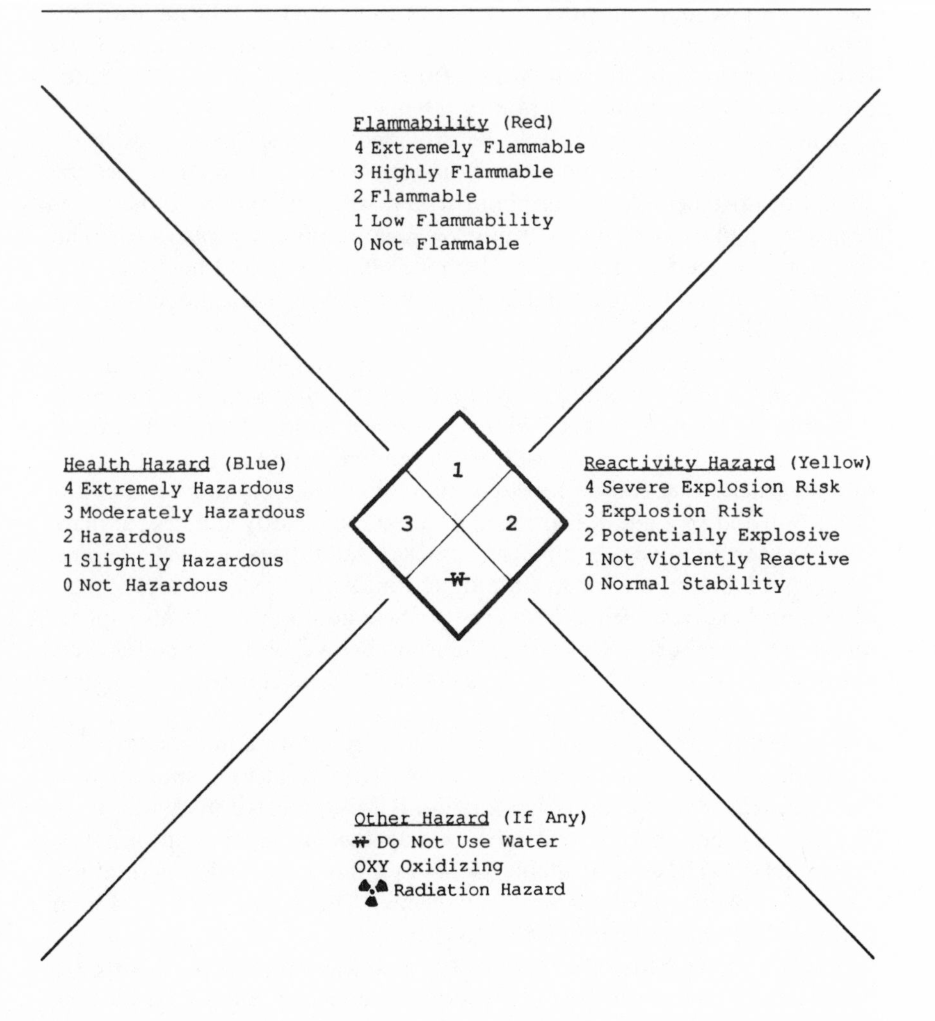

Other potential sources of information include persons involved in the emergency incident and observations of witnesses regarding the behavior, color, and odor of materials. Phone numbers on shipping papers can enable first responders to gather valuable information about materials in transit. Also, many local fire departments maintain lists of hazardous materials that are frequently transported through their area, and this information may aid in the identification process.

Emergency information systems can provide hazard information and guidance when supplied with the name of the material and the nature of the problem. Additionally, they can help with product identification. These information systems are discussed further in relation to notification procedures, which is the next priority after identification.

Immediate notification should be provided to those individuals in the vicinity who may be affected by the situation. Other notification procedures should be documented in a telephone roster, which is updated and checked regularly, and is part of a facility or area contingency plan. Alternate telephone numbers and 24-hour telephone numbers should be listed when available. Table 11.3 is a sample of a comprehensive telephone roster as prepared by FEMA.

The need for preplanning is obvious upon studying this table. Not only is it necessary to notify appropriate governmental agencies, it is extremely desirable to identify and be able to contact entities that can provide additional aid. These groups will vary depending upon locale and the types of emergencies that can be foreseen. In addition to local providers of fire protection and first aid, it is often appropriate to identify sources for items such as earth-moving equipment, trucks, sand bags, gravel, booms, skimmers, barriers, and neutralization chemicals. If there is a local source of technical experts, such as a university, these numbers should also appear on the telephone roster. Not every phone number will be contacted for every emergency, but it is far better to have the phone number listed and not need it than the other way around.

Included in Table 11.3 are various emergency information systems. The Chemical Transportation Emergency Center (CHEMTREC), sponsored by the Chemical Manufacturers Association (CMA), located in Washington, D.C., is the best known and most widely used emergency information center. CHEMTREC is available 24 hours a day to provide assistance to anyone involved in a chemical emergency. The CHEMTREC 24-hour hotline telephone number is 1-800-424-9300.

Several specialized information systems can be accessed through CHEMTREC. These include the Chlorine Emergency Plan (CHLOREP), Interagency Radiological Assistance Plan (IRAP), Transportation Emergency Reporting Procedure (TERP), Hazardous Emergency Leak Procedure (HELP), and the Pesticides Safety Team Network (PSTN). CHLOREP is a system which was developed by the Chlorine Institute and

Table 11.3
Sample Telephone Roster for Emergency Responses

Community Assistance:
- -Police
- -Fire
- -Civil Defense
- -Environmental Protection Agency
- -Department of Transportation
- -Public Works:
  - Water Supply
  - Sanitation
- -Rescue Squad
- -Ambulance
- -Hospitals
- -Utilities:
  - Gas
  - Phone
  - Electricity

Community Officials:
- -Mayor
- -City Manager
- -County Executive

Response Personnel:
- -On-Scene Coordinator
- -Agency Coordinators
- -Response Team Members

Bordering Political Regions:
- -Cities
- -Counties
- -River Basin Authorities
- -Interstate Compact
- -Regional Authorities
- -States

Industry:
- -Transporters
- -Chemical Producers/Consumers
- -Spill Cooperatives
- -Spill Response Teams

Volunteer Groups:
- -Red Cross
- -Salvation Army
- -Church Groups
- -Ham Radio Operators

Media:
- -Television
- -Newspaper
- -Radio

State Assistance:
- -Environmental Protection Agency
- -Civil Defense
- -Department of Transportation
- -Police
- -Public Health Dept.
- -Military Department

Federal Assistance:
- -U.S. Department of Transportation
- -U.S. Environmental Protection Agency
- -Federal Emergency Management Agency
- -National Response Center
- -Center for Disease Control (Contact via State/Local Public Health)
- -Emergency Response
- -Radioactive Material
- -U.S. Army
- -Nuclear Regulatory Commission

Other Emergency Assistance:
- -CHEMTREC
- -CHLOREP
- -NACA Pesticide Safety Team
- -Bureau of Explosives
- -Cleanup Contractor
- -Bomb Disposal and/or Explosive Ordinance Team, U.S. Army
- -Poison Control Center

chlorine manufacturers to deal with chlorine releases. IRAP provides guidance in dealing with emergencies involving radioactive materials. TERP and HELP were developed by Du Pont and Union Carbide, respectively, and provide emergency response information on each of the company's products. When a hazardous materials incident involves a pesticide, CHEMTREC will notify PSTN, which is a mutual aid system which was developed by the National Agricultural Chemicals Association.

The value of the guidance which can be obtained from an emergency information system is directly proportional to the amount of accurate information provided about the emergency situation. The following is a list of commonly required information:

Name of Caller
Callback Number
Materials Involved
Nature of the Problem
Location of the Problem
Name of Shipper or Manufacturer
Type of Container
Transport Vehicle Identification
Weather Conditions

In some emergencies, it may be necessary to evacuate an area. Such a step should never be undertaken lightly: In certain emergency situations, deaths have occurred that were attributable solely to evacuation. Removing people from their homes is never a simple matter. Complications likely to be encountered include the sick, the elderly, those on continual medication, separation of family members, and abandonment of pets and livestock—not to mention the basic human instinctive resistance to leaving homes and valuable possessions.

Initial technical questions in any evacuation involve the size of the area to be evacuated and the duration of the evacuation. For most incidents, many experts feel a one-mile radius is a reasonable goal to start with, although safety requirements will vary with each individual situation. The duration of the evacuation will depend on how competently the first phases of cleanup are managed.

Crowd control is closely related to evacuation. While some mechanical aids such as bullhorns and blinking lights are helpful, calmness, conciseness, and consistency on the part of those giving directions are necessities. Smoking on the part of the evacuees should be eliminated as much as possible; those involved in directing and carrying out the emergency response efforts should never smoke on the scene.

A list and count of all persons evacuated should be maintained. Those to be evacuated should be directed to one designated place where they can be convened to receive further directions. Local structures which can serve as emergency shelters or as staging areas for providing further directions should be identified in a contingency plan. Common examples are schools, warehouses, armories, and theaters.

Once an evacuation has been conducted, there will be the problem of maintaining control of the evacuated area. There will always be a few people who must (or feel they must) enter an evacuated area. For any incident longer than one to two days, it is recommended that a system involving passes, sign-in/sign-out procedures, and escorts be implemented.

With most spills, containment is the first response priority in order to keep the spill from being spread until remedial measures can be taken. Containment may require plugging a hole in a container, spreading absorbent material to soak up spilled chemicals, or erecting a barrier and closing off an entire area. A simplified hazardous materials spill response diagram is shown in Figure 11.3.

Depending on the type of material spilled, another concern may be vapor control. Vapors of certain materials may be toxic, others may be flammable. Two methods of vapor control are covering the liquid with

Figure 11.3
Simplified Hazardous Materials Spill Response Diagram

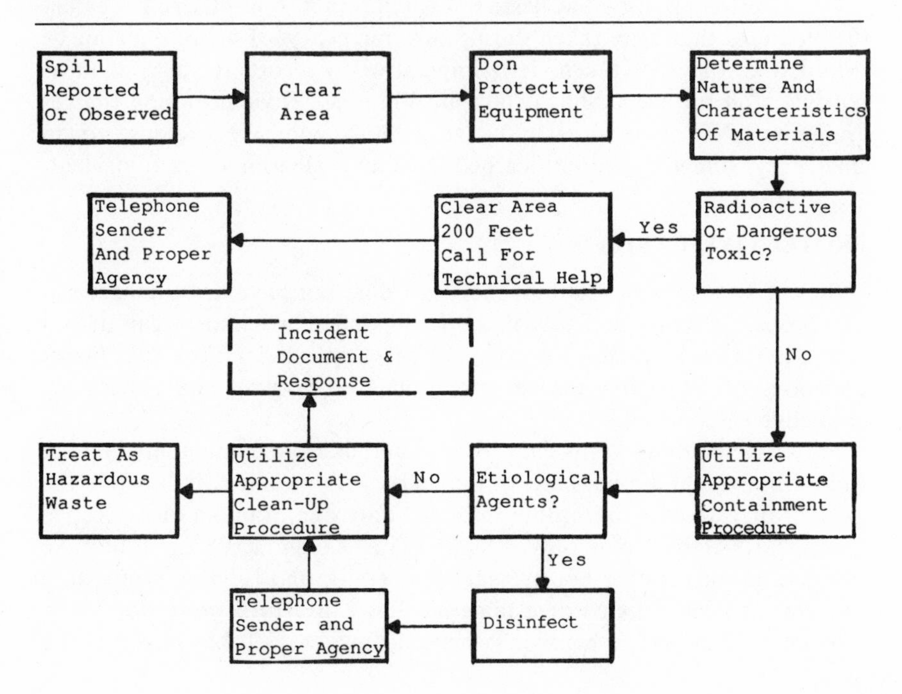

foam or cooling the liquid to lower the vapor pressure. In the first case, high expansion foams are recommended for long-term applications. In the second case, dry ice is usually the coolant of choice.

After immediate containment and vapor control, if necessary, the ultimate accumulation and disposal of the spilled material must be controlled. While cleanup is being implemented, the materials must be kept from entering groundwater supplies, sewer systems, and surface waters.

Providing the appropriate fire protection requires a knowledge of locally available equipment and staff. At least some generic identification of the burning substance is necessary before it can be countered with fire extinguishment. Figure 11.4 depicts appropriate fire extinguishers for various types of fires.

There are a growing number of fires and potential explosives where the application of water would be disastrous. Pressurized containers cannot be quenched with water without risk of explosion. In some cases, it is better to let the material burn, particularly if the products of combustion are less hazardous than the original material. This is often the case with fires involving pesticides.

Due to the increasing complexity of hazardous materials incidents, it is recommended that fire protection efforts be considered an important, but interdisciplinary component of emergency response. Firefighters should coordinate with the OSC and base their actions on information gathered from all appropriate technical resources.

Once a crisis situation has passed, it is important to be able to document the measures that were taken during the emergency. If actions cannot be recorded as they are taken, it is imperative to record them as soon as possible after the event while details are still fresh. Documentation enables evaluation. No matter how effective or how poor the response to an emergency, something can be learned from an evaluation of each incident.

PROTECTIVE DEVICES

Close contact with hazardous materials does not have to be dangerous. The proper selection and use of the right kind of equipment can afford protection against virtually every kind of chemical risk. The two broad categories of protective devices are protective clothing and respiratory protective equipment.

Protective clothing keeps dangerous materials from coming into contact with skin, eyes, and other parts of the body. The "moon suits" used in the cleanup of extremely hazardous sites are only one example of protective equipment. Many commonly available items can serve as protective clothing: gloves, rubber boots, goggles, overalls, and rubber aprons. It is important to select items of a material that will afford protection when exposed to the specific hazardous material involved. Table 11.4 lists the

Figure 11.4
Appropriate Fire Extinguishers for Various Types of Fires

Fire Extinguishers			
Type of Fire → / **Extinguisher to Use** ↓	A **CLASS A** • Ordinary Combustibles • Wood • Paper • Cloth	B **CLASS B** • Flammable liquids, grease • Gasoline • Paints • Oils	C **CLASS C** • Energized electrical equipment • Motors • Switches • Fuse boxes
Multi-Purpose Dry Chemical • Stored Pressure • Cartridge Operated	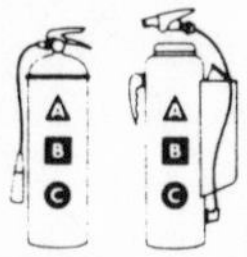 A B C / A B C	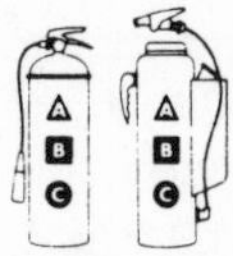 A B C / A B C	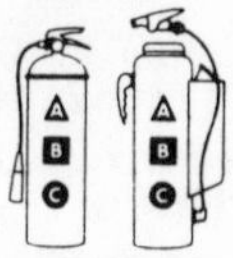 A B C / A B C
Ordinary Dry Chemical • Stored Pressure • Cartridge Operated	**No**	 B C / B C	 B C / B C
Carbon Dioxide • Self-Propelling	**No**	 B C	 B C
Water • Stored Pressure • Pump Tank	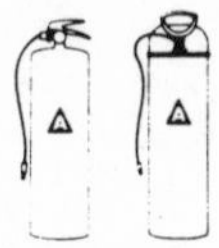 A / A	**No**	**No**
AFFF Foam • Stored Pressure	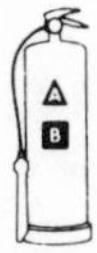 A B	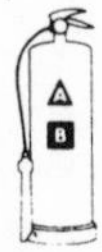 A B	**No**
Halon 1211 • Stored Pressure	 A B C	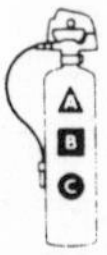 A B C	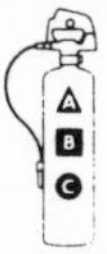 A B C
D **CLASS D** Involve Combustible Metals	Combustible Metals Include: • Magnesium • Titanium • Zirconium • Sodium • Potassium • Lithium • Dry Powder Aluminum • Calcium • Thorium • Sodium-Potassium	 Met-L-X Chemical	 Cartridge operated

Table 11.4
Protective Clothing Materials Rated by Chemical

Chemical	Chemical Makeup of Clothing			
	Butyl Rubber	Polyvinyl Chloride	Neoprene	Natural Rubber
Acetaldehyde	E	U	E	G
Acetic Acid	E	E	E	E
Acetone	E	P	G	E
Acrylenitrile	G	U	G	F
Ammonium Hydroxide	E	E	E	E
Amyl Acetate	F	P	F	P
Amyl Alcohol	E	E	E	E
Ammonia	E	E	G	E
Aniline	F	F	G	F
Benzaldehyde	G	U	F	F
Benzene	P	F	P	P
Butyl Acetate	F	P	F	G
Butyl Alcohol	E	E	E	E
Calcium Chloride	U	E	U	U
Calcium Hydroxide	E	E	E	E
Carbon Tetra-Chloride	P	U	F	P
Castor oil	F	G	F	P
Chlorine	G	U	G	F
Citric Acid	E	E	E	E
Cottonseed oil	F	G	G	G
Cresol	G	U	G	G
Dibutyl Phthalate	G	P	G	P
Diesel fuel	P	U	G	P
Diethanolamine	E	E	E	G
Ethyl Acetate	G	P	G	F
Ethyl Alcohol	E	E	E	E
Ethyl Ether	E	G	E	G
Ethylene Glycol	E	E	E	E
Formaldehyde	E	G	E	E
Formic Acid	E	E	E	E
Furfural	G	F	G	G
Gasoline (leaded)	P	U	G	P
Glycerine	E	E	E	E
Hexane	P	F	F	P
Hydrazine	G	U	F	G
Hydrochloric Acid	G	E	E	G
Hydrofluric Acid	G	E	E	G
Hydrogen Peroxide(30%)	G	G	G	G
Isopropyl Alcohol	E	E	E	E

Table 11.4 (Continued)

Chemical	Chemical Makeup of Clothing			
	Butyl Rubber	Polyvinyl Chloride	Neoprene	Natural Rubber
Kerosene	F	F	E	F
Methyl Alcohol	E	E	E	E
Methyl Chloride	P	G	P	P
Methyl Ethyl Ketone	E	P	G	G
Methylene Chloride	G	P	G	F
Methyl Isobutyl Ketone	E	P	F	F
Mineral oils	F	G	E	F
Morpholine	F	P	E	E
Naphthas, Aliphatic	F	F	E	F
Nitric Acid	F	E	G	F
Nitric Acid, Fuming	P	G	P	P
Oxalic Acid	E	U	E	E
Phenol	G	E	E	F
Phosphoric Acid	E	E	E	G
Potassium Dichromate	E	E	F	F
Potassium Hydroxide	E	E	E	E
Sodium Hydroxide	E	E	E	E
Sulfuric Acid	G	E	G	G
Tetraethyl Lead	G	U	E	F
Toluene	P	P	F	P
Trichloroethylene	P	U	F	F
Tricresyl Phosphate	F	P	G	F
Triethanolamine	G	G	E	G
Turpentine	F	F	G	F
Vegetable oils	G	G	E	G
Zinc Chloride	E	U	E	E

Legend: E - Excellent
G - Good
F - Fair
P - Poor
U - Unsatisfactory

effectiveness of protective clothing materials in dealing with various chemicals.

Respiratory protective devices may be self-contained, supplying air through a mask and hose. Others are designed to filter out contaminants by various methods. Selection of the appropriate type depends on the material presenting the inhalation hazard and the length of exposure. Short of a totally self-contained body suit, the best respiratory protection is afforded by full-face respirators with supplied air. For full effectiveness, these respirators must be face fit to the individual using them.

It is appropriate to address the phenomenon of overprotection in a discussion of protective equipment. The normal response in an emergency situation is to provide maximum protection for all workers. However, protective clothing and breathing devices can be cumbersome and confining, tiring workers and making it difficult to complete certain tasks. Workers who are outfitted unnecessarily may also be tempted to discard protective equipment in the next emergency. Although it is desirable to err on the side of caution, the risk of excess must be considered when protective clothing and breathing devices are used.

Another topic related to the use of protective equipment is decontamination—that is, the cleanup after the cleanup. In any emergency response, personnel and equipment are likely to become contaminated. Decontamination usually involves some type of washing procedure, which may in turn generate contaminated runoff and spent wash solutions. It is imperative that all contaminated by-products be dealt with properly, even those that are not generated by the primary clean-up activities. Contaminated equipment, which cannot be decontaminated, must be disposed of properly.

PREVENTION

Obviously, prevention is preferable to any emergency response, no matter how well executed. The major constituents of any prevention plan are safety and training, which are considered essential industrial work practices. Closely related is community awareness. A hazardous materials incident may result in very little damage to human health and the environment if it occurs in a community with a high level of disaster preparedness. Such preparation not only enables correct responses to a given situation, but also lessens the possibility of community or facilitywide panics.

Panic in an emergency situation can result in as many, if not more, detrimental effects as the hazardous material itself. Regardless of the number of experts available at various federal and state agencies, the front line is often the fire chief or local sheriff whose staff has just responded to an accident. A trained complement of local public employees and an informed community can prevent accidents from becoming tragedies.

12

REGULATORY COMPLIANCE AND WORK PRACTICES

Assuring better compliance over time and getting firms to affirmatively manage their environmental impacts, instead of reacting to crises, are major concerns.

—Michael Levin
Chief, Regulatory Reform Staff, EPA

Regulatory compliance is a significant aspect of conducting business today. The complex scheme of obligations posed by environmental legislation represents two costs—the effort and expenditure required to achieve compliance, and the fines, penalties, and liabilities that may be incurred as a result of noncompliance. It should be noted that ignorance of regulatory requirements is viewed by federal, state, and local governments as no excuse for noncompliance.

Compliance is achieved through work practices that are mandated by management. However, compliance is not the only possible goal of a hazardous materials management program, as discussed in Chapter 9. Similarly, technical regulatory compliance is not the only possible goal of appropriate work practices. Attitudes and approaches that generally help to avoid or minimize conflict and litigation are highly desirable.

DEFINING COMPLIANCE

Compliance criteria are normally found in two types of documents: applicable environmental laws and facility-specific permit conditions. Other potential sources of compliance criteria include court actions, consent decrees, contracts, leases, and labor agreements. Typical compliance requirements address monitoring procedures; control techniques; schedules; record-keeping requirements; reporting and notification requirements; accepted disposal, treatment, storage, or shipping methods; emergency response procedures; employee training; and inspections.

The definition of compliance requirements is not a one-time effort. Environmental regulations are continuously promulgated, revised, challenged, and interpreted. A well-thought-out, properly targeted legislative and regulatory strategy includes remaining current with all applicable policy developments. This is especially important for facilities with unique or unusual concerns.

It is highly desirable to anticipate, rather than simply react to, regulatory requirements. Perhaps the biggest mistake managers make in defining compliance is to take proposed permit conditions and general rules at face value, ignoring the details that often address exceptions, exclusions, and exemptions. Permitting and rule making both offer opportunities for negotiation and constructive communication with regulatory agencies. Discerning managers recognize and use all of the opportunities open to them for minimizing actual or potential conflict with regulatory agencies at each point of the regulatory relationship.

ENVIRONMENTAL AUDITS

Environmental audits have become a widely accepted work practice and are intended to aid in achieving regulatory compliance. Since there are no federal or regulatory provisions defining environmental auditing, it can and does mean different things in various circumstances and contexts as applied to specific industrial activities.

In its broadest form, the environmental audit is a vehicle to help management identify environmental problems with potential adverse effects and avoid or minimize liability. The audit can also point out requirements for changes in operating or administrative procedures to improve systems for compliance, quality assurance, and internal reporting.

Implementing an audit requires the selection of an audit team, identification of regulatory requirements, and determination of required data. Data can be collected in a number of ways, including telephone and direct interviews, written questionnaires, site inspections, and searches of plant operating and analytical records. It is necessary to know as accurately as possible the quantities and characteristics of the raw materials entering the facility, and the quantities, characteristics, and final disposition of each waste stream. Often the audit involves a two-step procedure: the administrative audit, which tracks the historical records of the movement of hazardous materials through a facility, and a field audit, which documents current hazardous materials handling practices.

One of the most important outputs of an audit is a compliance profile. The profile should state in detail the specific applicable regulatory requirements, the conditions that are required for compliance, and the extent to which these conditions are being met. The final analysis of information gathered by an environmental audit should address the following topics:

1. Areas in which existing practices are adequate;
2. Areas in which existing practices are inadequate;
3. Previously unaddressed compliance issues; and
4. Potential future problem areas.

The audit can then be used as a guide in the development of efficient and effective programs which will ensure compliance and minimize both corporate liability and the potential release of hazardous materials.

PREPARING FOR AN INSPECTION

Compliance is often determined by the results of an agency inspection. Certain categories of facilities are more likely to be inspected than others, but no operation which involves hazardous materials is exempt from inspections. It is important to consider inspection-related issues and develop appropriate responses *before* the event.

EPA and its contractors are granted broad rights to inspect under various environmental laws. If public access to a facility is ordinarily restricted, the owners can usually insist on a search warrant before granting access. However, this is seldom recommended as a course of action, unless the delay is deemed valuable enough to warrant possibly damaging the relationship with the regulatory agency. In any case, the facility owner/operator retains the right to check and verify the inspector's credentials. The inspector should also be willing to explain the purpose of the visit, and provide a copy of any employee complaint that may be involved (with the employee's name deleted, if the employee has requested anonymity).

It is also reasonable to ask which areas of the facility are to be inspected. The inspection can then be limited to the areas originally specified. A company official should accompany the inspector at all times.

It is desirable that pertinent records and permits be available for easy access. Excess documents which are not directly related to the objectives of the inspection should not be provided to the inspector.

If the inspector takes any samples, duplicate samples should be taken at the same time, using the same sampling technique. This affords the opportunity to refute or verify the agency's findings based on independent analyses.

Finally, a copy of the agency's preliminary inspection form should be requested along with an indication of any violations or deficiencies. Also, a copy of the final report should be requested as soon as it is available.

NONLITIGATION DISPUTE RESOLUTION

The avoidance of litigation provides a great incentive for industry to develop methods to police its own environmental practices. The terms and conditions contained in permits may be every bit as important, in terms of profitability, as a major corporate contract. If compliance cannot be or is not maintained, significant costs will result in the long term.

Environmental auditing, as previously discussed, has been endorsed by various groups including EPA as a nonconfrontational vehicle for ensuring regulatory compliance. There are also numerous dispute avoidance strategies associated with permitting processes, all of which involve some type of negotiation as opposed to confrontation.

These strategies, and other forms of arbitration, can significantly reduce the scope of conflicts between regulators and members of the regulated community. It is important to identify the areas in any pertinent regulation that leave room for negotiation and match negotiation strategies as closely as possible with corporate long-term interests and goals. It is preferable to confront what may seem to be an unreasonable requirement with a suggested alternative rather than to declare an impasse.

In all such negotiations, the burden is on the permit applicant to research and justify proposed deviations from standard agency permitting policies. Review of any proposal by agency staff can always be requested, whether or not informal acceptance is legally binding. A written indication of the results of an agency review can be of great future benefit. At a minimum, if a proposal is not acceptable to an agency, it is preferable to find out before money has been spent on implementation. The chances of successful negotiation are far greater when the process is initiated, pursued, and documented by the applicant, rather than the other way around.

Cost effective management requires a spirit of cooperation in order to avoid the costs, delays, and frustration of litigation. There is no perfect solution to the concerns pertaining to hazardous materials management. A partial solution is vastly superior to no solution.

ONGOING REGULATORY AGENCY CONTACT

It is highly recommended that contact with regulatory agencies be initiated before the need arises to resolve a dispute. Routine periodic contact with regulatory officials serves to strengthen the lines of communication. Items to be discussed during such contact might include the status of current regulations, upcoming draft or interim regulations, and changes in enforcement or permitting policies.

Good work practice generally involves contacting the regulatory agencies at least quarterly, and more often if site-specific or facility-specific conditions warrant.

13

SOURCES OF ADDITIONAL INFORMATION

Knowledge is of two kinds. We know a subject ourselves, or we know where we can find information upon it.

—Samuel Johnson, 1775

TOLL-FREE NUMBERS

Consumer Product Safety Commission Hotlines

Maryland

1-800-492-8363

Alaska, Hawaii, Puerto Rico, and Virgin Islands

1-800-638-8333

Continental U.S.

1-800-638-2772

EPA RCRA/SUPERFUND Hotline

1-800-424-9346

EPA SPWA Hotline

1-800-426-4791

EPA Chemical Emergency Preparedness Program Hotline
(regulatory interpretation regarding emergency preparedness requirements)

1-800-535-0202

National Response Center

1-800-424-8802 (for emergency use only)

CHEMTREC

1-800-424-9300 (for emergency use only)

National Pesticide Telecommunication Network

1-800-858-PEST

Chemical Manufacturers Association Chemical Referral Center
(information referral to manufacturers of chemical products)
1-800-CMA-8200

National Resources Defense Council Toxics Line
(health effects of chemicals)
1-800-648-NRDC

OTHER NUMBERS

Department of Transportation
202-426-2311

Nuclear Regulatory Commission
202-492-7000

Occupational Safety and Health Administration
202-523-8148

EPA Central Office
Regarding:
Asbestos
202-559-1909
Clean Air Act
202-382-7400
Clean Water Act
202-475-9545
Toxic Substance Control Act
202-554-1404

EPA REGIONAL OFFICES

Region I:
617-223-7210
Connecticut
Maine
Massachusetts
New Hampshire
Rhode Island
Vermont

Region II:
212-264-2525
New Jersey
New York
Puerto Rico
Virgin Islands

Region III:
215-597-9814
Delaware
District of Columbia
Maryland
Pennsylvania
Virginia
West Virginia

Region IV:
404-881-4727
Alabama
Florida
Georgia
Kentucky
Mississippi
North Carolina
South Carolina
Tennessee

Region V:
312-353-2000
Illinois
Indiana
Michigan
Minnesota
Ohio
Wisconsin

Region VI:
214-767-2600
Arkansas
Louisiana
New Mexico
Oklahoma
Texas

Region VII:
816-374-5493
Iowa
Kansas
Missouri
Nebraska

Region VIII:
303-837-3895
Colorado
Montana
North Dakota
South Dakota
Utah
Wyoming

Region IX:
415-974-8153
Arizona
California
Hawaii
Nevada
American Samoa
Guam

Region X:
206-442-1220
Alaska
Idaho
Oregon
Washington

APPENDIX: SUPPLEMENTARY TECHNICAL INFORMATION

Table A.1
Priority Pollutants*

1	*Acenaphthene
2	*Acrolein
3	*Acrylonitrile
4	*Benzene
5	*Benzidine
6	*Carbon tetrachloride (tetrachloromenthane)
	*chlorinated benzenes (other than dichlorobenzenes)
7	Chlorobenzene
8	1,2,4 trichlorobenzene
9	Hexachlorobenzene
	*chlorinated ethanes (including 1,2-dichloethane, 1,1,1-trichloroethane and hexachloroethane)
10	1,2-dichloroethane
11	1,1,1-trichloroethane
12	Hexachloroethane
13	1,1-dichloroethane
14	1,1,2-trichloroethane
15	1,1,2,2-tetrachloroethane
16	Chloroethane *chloroalkyl ethers (chloromethyl, chloroethyl & mixed ethers)
17	bis(chloromethyl) ether**
18	Bis(2-choloroethyl) ether
19	2-chloroethyl vinyl ether (mixed) *chlorinated naphthalene
20	2-chloronaphthalene
	*chlorinated phenols (other than those listed elsewhere; includes trichlorophenols and chlorinated cresols)
21	2,4,6-trichlorophenol
22	Parachlorometa cresol
23	*Chloroform (trichloromethane)
24	*2-chlorophenol
	*dichlorobenzenes
25	1,2-dichlorobenzene
26	1,3-dichlorobenzene
27	1,4-dichlorobenzene
	*dichlorobenzidine
28	3,3′-dichlorobenzidine
	*dichloroethylenes (1,1-dichloroethylene and 1,2-dichloroethylene
29	1,1-dichloroethylene
30	1,2-trans-dichloroethylene
31	*2,4-dichlorophenol
	*dichloropropane and dichloropropene
32	1,2-dichloropropane
33	1,2-dichloropropylene (1, 3-dichloropropene)
34	*2,4-dimethylphenol
	*dinitrotoluene
35	2,4-dinitrotoluene
36	2,6-dinitrotoluene
37	*1,2-diphenylhydrazine
38	*Ethylbenzene
39	*Fluoranthene *haloethers (other than those listed elsewhere)
40	4-chlorophenyl phenyl ether
41	4-bromophenyl phenyl ether
42	Bis(2-chloroisopropyl) ether
43	Bis(2-chloroethoxy) methane *halomethanes (other than those listed elsewhere)
44	Methylene chloride (dichloromethane)
45	Methyl chloride (chloromethane)
46	Methyl bromide (bromomethane
47	Bromoform (tribromomethane

48 Dichlorobromomethane
49 Trichlorofluoromethane**
50 Dichlorodifluoromethane**
51 Chlorodibromomethane
52 ***Hexachlorobutadiene**
53 ***Hexachlorocyclopentadiene**
54 ***Isophorone**
55 ***Naphthalene**
56 ***Nitrobenzene *nitrophenols** (including 2,4-dinitrophenol and dinitrocresol)
57 2-Nitrophenol
58 4-Nitrophenol
59 ***2,4-dinitrophenol**
60 4,6-dinitro-o-cresol ***nitrosamines**
61 N-nitrosodimethylamine
62 N-nitrosodiphenylamine
63 N-nitrosodi-n-propylamine
64 ***Pentachlorophenol**
65 ***Phenol *phthalate esters**
66 Bis(2-ethylhexyl) phthalate
67 Butyl benzyl phthalate
68 Di-n-butyl phthalate
69 Di-n-octyl phthalate
70 Diethyl phthalate
71 Dimethyl phthalate ***polynuclear aromatic hydrocarbons**
72 Benzo(a)anthracene (1,2-benzanthracene)
73 Benzo(a)pyrene (3,4-benzopyrene)
74 3,4-benzofluoranthene
75 Benzo(k)fluoranthane (11,12-benzofluoranthene)
76 Chrysene
77 Acenaphthylene
78 Anthracene
79 Benzo(ghi)perylene (1,12-benzoperylene)
80 Fluorene
81 Phenanthrene
82 Dibenzo(a,h)anthracene (1,2,5,6-dibenzanthracene)
83 Indeno (1,2,3-cd)pyrene (2,3-o-phenylenepyrene)
84 Pyrene
85 ***Tetrachloroethylene**
86 ***Toluene**
87 ***Trichloroethylene**
88 ***Vinyl chloride** (chloro-ethylene) pesticides and metabolites
89 ***Aldrin**
90 ***Dieldrin**
91 ***Chlordane** (technical mixture and metabolites) DDT and metabolites
92 4,4´-DDT
93 4,4´-DDE (p,p´-DDX)
94 4,4´-DDD (p,p´-TDE) ***endosulfan & metabolites**
95 A-endosulfan-Alpha
96 B-endosulfan-Beta
97 Endosulfan sulfate ***endrin and metabolites**
98 Endrin
99 Endrin aldehyde ***heptachlor and metabolites**
100 Heptachlor
101 Heptachlor epoxide ***hexachlorocyclohexane** (all isomers)
102 a-BHC-Alpha
103 b-BHC-Beta
104 r-BHC-(lindane)-Gamma
105 g-BCH-Delta ***polychlorinated biphenyls (PCB´s)**
106 PCB-1242 (Arochlor 1242)
107 PCB-1254 (Arochlor 1254)
108 PCB-1221 (Arochlor 1221)

Table A.1 (Continued)

109	PCB-1232 (Arochlor 1232)
110	PCB-1248 (Arochlor 1248)
111	PCB-1260 (Arochlor 1260
112	PCB-1016 (Arochlor 1016)
113	***Toxaphene**
114	***Antimony** (total)
115	***Arsenic** (total)
116	***Asbestos** (fibrous)
117	***Beryllium** (total)
118	***Cadmium** (total)
119	***Chromium** (total)
120	Copper (total)
121	***Cyanide** (total)
122	***Lead** (total)
123	***Mercury** (total)
124	***Nickel** (total)
125	***Selenium** (total)
126	***Silver** (total)
127	***Thallium** (total)
128	***Zinc** (total)
129	2,3,7,8-tetrachlorodi-benzo-p-dioxin (TCDD)

* The original list consisted of 65 specific compounds and chemical classes, indicated in bold type. When various forms of certain categories were broken out, the original list included 13 metals, 114 organic chemicals, asbestos, and cyanide. Three organic chemicals were deleted in 1981.

** Deleted in 1981.

Table A.2
Reportable Quantities (RQs)

Material	RQ in Pounds	Material	RQ in Pounds
Acetaldehyde	1000	Antimony pentachloride	1000
Acetic acid	5000	Antimony potassium tartrate	100
Acetic anhydride	5000	Antimony tribromide	1000
Acetone cyanohydrin	10	Antimony trichloride	1000
Acetyl bromide	5000	Antimony trifluoride	1000
Acetyl chloride	5000	Antimony trioxide	1000
Acrolein	1	Arsenic disulfide	5000
Acrylonitrile	100	Arsenic pentoxide	5000
Adipic acid	5000	Arsenic trichloride	5000
Aldrin	1	Arsenic trioxide	5000
Allyl alcohol	100	Arsenic trisulfide	5000
Allyl chloride	1000	Barium cyanide	10
Aluminum sulfate	5000	Benzene	1000
Ammonia	100	Benzoic acid	5000
Ammonium acetate	5000	Benzonitrile	5000
Ammonium benzoate	5000	Benzoyl chloride	1000
Ammonium bicarbonate	5000	Benzyl chloride	100
Ammonium bichromate	1000	Beryllium chloride	5000
Ammonium bifluoride	100	Beryllium fluoride	5000
Ammonium bisulfite	5000	Beryllium nitrate	5000
Ammonium carbamate	5000	Butyl acetate	5000
Ammonium carbonate	5000	Butylamine	1000
Ammonium chloride	5000	n-Butyl phthalate	10
Ammonium chromate	1000	Butyric acid	5000
Ammonium citrate	5000	Cadmium acetate	100
Ammonium fluoborate	5000	Cadmium bromide	100
Ammonium fluoride	100	Cadmium chloride	100
Ammonium hydroxide	1000	Calcium arsenate	1000
Ammonium oxalate	5000	Calcium arsenite	1000
Ammonium silicofluoride	1000	Calcium carbide	10
Ammonium sulfamate	5000	Calcium chromate	1000
Ammonium sulfide	100	Calcium cyanide	10
Ammonium sulfite	5000	Calcium dodecylbenzenesulfonate	1000
Ammonium tartrate	5000	Calcium hypochlorite	10
Ammonium thiocyanate	5000	Captan	10
Ammonium thiosulfate	5000	Carbaryl	100
Amyl acetate	5000	Carbofuran	10
Aniline	5000	Carbon disulfide	100

Table A.2 (Continued)

Material	RQ in Pounds	Material	RQ in Pounds
Carbon tetrachloride	5000	Dichloropropene-Dichloropropane mixture	100
Chlordane	1	2,2-Dichloropropionic acid	5000
Chlorine	10	Dichlorvos	10
Chlorobenzene	100	Dieldrin	1
Chloroform	5000	Diethylamine	100
Chlorosulfonic acid	1000	Dimethylamine	1000
Chlorpyrifos	1	Dinitrobenzene	100
Chromic acetate	1000	Dinitrophenol	10
Chromic acid	1000	Dinitrotoluene	1000
Chromic sulfate	1000	Diquat	1000
Chromous chloride	1000	Disulfoton	1
Cobaltous bromide	1000	Diuron	100
Cobaltous formate	1000	Dedecylbenzenesulfonic acid	1000
Cobaltous sulfamate	1000	Endosulfan	1
Coumaphos	10	Endrin	1
Cresol	1000	Epichlorohydrin	1000
Crotonaldehyde	100	Ethion	10
Cupric acetate	100	Ethylbenzene	1000
Cupric acetoarsenite	100	Ethylenediamine	5000
Cupric chloride	10	Ethylene dibromide	1000
Cupric nitrate	100	Ethylene dichloride	5000
Cupric oxalate	100	EDTA	5000
Cupric sulfate	10	Ferric ammonium citrate	1000
Cupric sulfate ammoniated	100	Ferric ammonium oxalate	1000
Cupric tartrate	100	Ferric chloride	1000
Cyanogen chloride	10	Ferric fluoride	100
Cyclohexane	1000	Ferric nitrate	1000
2,4-D Acid	100	Ferric sulfate	1000
2,4-D Esters	100	Ferrous ammonium sulfate	1000
DDT	1	Ferrous chloride	100
Diazinon	1	Ferrous sulfate	1000
Dicamba	1000	Formaldehyde	1000
Dichlobenil	100	Formic acid	5000
Dichlone	1	Fumaric acid	5000
Dichlorobenzene	100	Furfural	5000
Dichloropropane	1000		
Dichloropropene	100		

Material	RQ in Pounds	Material	RQ in Pounds
Guthion	1	Methyl parathion	100
Heptachlor	1	Mevinphos	10
Hexachlorocyclopentadiene	1	Mexacarbate	1000
Hydrochloric acid	5000	Monoethylamine	100
Hydrofluoric acid	100	Monomethylamine	100
Hydrogen cyanide	10	Naled	10
Hydrogen sulfide	100	Naphthalene	100
Isoprene	100	Naphthenic acid	100
Isopropanolamine dodecylbenzenesulfonate	1000	Nickel ammonium sulfate	5000
Kelthane	10	Nickel chloride	5000
Kepone	1	Nickel hydroxide	1000
Lead acetate	5000	Nickel nitrate	5000
Lead arsenate	5000	Nickel sulfate	5000
Lead chloride	100	Nitric acid	1000
Lead fluoborate	100	Nitrobenzene	1000
Lead fluoride	100	Nitrogen dioxide	10
Lead iodide	100	Nitrophenol	100
Lead nitrate	100	Nitrotoluene	1000
Lead stearate	5000	Paraformaldehyde	1000
Lead sulfate	100	Parathion	1
Lead sulfide	5000	Pentachlorophenol	10
Lead thiocyanate	100	Phenol	1000
Lindane	1	Phosgene	10
Lithium chromate	1000	Phosphoric acid	5000
Malathion	100	Phosphorus	1
Maleic acid	5000	Phosphorus oxychloride	1000
Maleic anhydride	5000	Phosphorus pentasulfide	100
Mercaptodimethur	10	Phosphorus trichloride	1000
Mercuric cyanide	1	Polychlorinated biphenyls	10
Mercuric nitrate	10	Potassium arsenate	1000
Mercuric sulfate	10	Potassium arsenite	1000
Mercuric thiocyanate	10	Potassium bichromate	1000
Mercurous nitrate	10	Potassium chromate	1000
Methoxychlor	1	Potassium cyanide	10
Methyl mercaptan	100	Potassium hydroxide	1000
Methyl methacrylate	1000	Potassium permanganate	100
		Propargite	10
		Propionic acid	5000

Table A.2 (Continued)

Material	RQ in Pounds	Material	RQ in Pounds
Propionic anhydride	5000	Tetraethyl lead	10
Propylene oxide	100	Tetraethyl pyrophosphate	10
Pyrethrins	1	Thallium sulfate	100
Quinoline	5000	Toluene	1000
Resorcinol	5000	Toxaphene	1
Selenium oxide	10	Trichlorfon	100
Silver nitrate	1	Trichloroethylene	1000
Sodium	10	Trichlorophenol	10
Sodium arsenate	1000	Triethanolamine dodecylbenzenesulfonate	1000
Sodium arsenite	1000	Triethylamine	5000
Sodium bichromate	1000	Trimethylamine	100
Sodium bifluoride	100	Uranyl acetate	100
Sodium bisulfite	5000	Uranyl nitrate	100
Sodium chromate	1000	Vanadium pentoxide	1000
Sodium cyanide	10	Vanadyl sulfate	1000
Sodium dodecylbenzenesulfonate	1000	Vinyl acetate	5000
Sodium fluoride	1000	Vinylidene chloride	5000
Sodium hydrosulfide	5000	Xylene	1000
Sodium hydroxide	1000	Xylenol	1000
Sodium hypochlorite	100	Zinc acetate	1000
Sodium methylate	1000	Zinc ammonium chloride	1000
Sodium nitrite	100	Zinc borate	1000
Sodium phoshpate, dibasic	5000	Zinc bromide	1000
Sodium phosphate, tribasic	5000	Zinc carbonate	1000
Sodium selenite	100	Zinc chloride	1000
Strontium chromate	1000	Zinc cyanide	10
Strychnine	10	Zinc fluoride	1000
Styrene	1000	Zinc formate	1000
Sulfuric acid	1000	Zinc hydrosulfite	1000
Sulfur monochloride	1000	Zinc nitrate	1000
2,4,5-T acid	1000	Zinc phenolsulfonate	5000
2,4,5-T amines	5000	Zinc phosphide	100
2,4,5-T esters	1000	Zinc silicofluoride	5000
2,4,5-T salts	1000	Zinc sulfate	1000
TDE	1	Zirconium nitrate	5000
2,4,5-TP acid	100	Zirconium potassium fluoride	1000
2,4,5-TP acid esters	100	Zirconium sulfate	5000
		Zirconium tetrachloride	5000

Table A.3
Clean Water Act (1987) Funding

State	(%)*	Funding ($)
Alabama	1.13	203,562,000
Alaska	0.61	108,954,000
Arizona	0.68	122,958,000
Arkansas	0.66	119,088,000
California	7.23	1,301,994,000
Colorado	0.81	145,620,000
Connecticut	1.24	223,020,000
Delaware	0.50	89,370,000
Florida	3.41	614,502,000
Georgia	1.71	307,800,000
Hawaii	0.78	140,994,000
Idaho	0.50	89,370,000
Illinois	4.57	823,338,000
Indiana	2.44	438,732,000
Iowa	1.37	246,384,000
Kansas	0.91	164,322,000
Kentucky	1.29	231,696,000
Louisiana	1.11	200,124,000
Maine	0.78	140,922,000
Maryland	2.45	440,298,000
Massachusetts	3.43	618,084,000
Michigan	4.35	782,766,000
Minnesota	1.86	334,602,000
Mississippi	0.91	164,016,000
Missouri	2.80	504,666,000
Montana	0.50	89,370,000
Nebraska	0.52	93,114,000
Nevada	0.50	89,370,000
New Hampshire	1.01	181,926,000
New Jersey	4.13	743,922,000
New Mexico	0.50	89,370,000
New York	11.16	2,009,376,000
North Carolina	1.83	328,554,000
North Dakota	0.50	89,370,000
Ohio	5.69	1,024,848,000
Oklahoma	0.82	147,078,000
Oregon	1.16	205,650,000
Pennsylvania	4.01	721,116,000
Rhode Island	0.68	122,238,000

Table A.3 (Continued)

State	(%)*	Funding ($)
South Carolina	1.04	186,498,000
South Dakota	0.50	89,370,000
Tennessee	1.47	264,456,000
Texas	4.62	832,068,000
Utah	0.53	95,922,000
Vermont	0.50	89,370,000
Virginia	2.07	372,564,000
Washington	1.76	316,584,000
West Virginia	1.58	283,788,000
Wisconsin	2.73	492,156,000
Wyoming	0.50	89,370,000

* Other allocations are: District of Columbia, $89,370,000 American Samoa, $16,344,000; Guam, $11,826,000; Northern Marianas, $7,596,000; Puerto Rico, $237,438,000; Pacific Trust Territory, $23,310,000; Virgin Islands, $9,486,000.

Table A.4
Drinking Water Contaminants for which Standards Must Be Established by EPA*

Volatile Organic Chemicals:	
-Trichloroethylene	-Benzene
-Tetrachloroethylene	-Chlorobenzene
-Carbon tetrachloride	-Dichlorobenzene(s)
-1.1.1-Trichloroethane	-Trichlorobenzene(s)
-1,2-Dichloroethane	-trans-1,2-Dichloroethylene
-Vinyl chloride	-1,1-Dichloroethylene
-Methylene chloride	-cis-1,2-Dichloroethylene
Microbiology and Turbidity:	
-Total coliforms	-Viruses
-Turbidity	-Standard plate count
-Giardia lamblia	-Legionella
Inorganics:	
-Arsenic	-Molybdenum*
-Barium	-Asbestos
-Cadmium	-Sulfate
-Chromium	-Copper
-Lead	-Vanadium*
-Mercury	-Sodium*
-Nitrate	-Nickel
-Selenium	-Zinc*
-Silver*	-Thallium
-Fluoride	-Beryllium
-Aluminum*	-Cyanide
-Antimony	
Synthetic Organics:	
-Endrin	-1,1,2-Trichloroethane
-Lindane	-Vydate
-Methoxychlor	-Simazine
-Toxaphene	-Polynuclear Aromatic Hydrocarbons
-2,4-D	-Polychlorinated Biphenyls
-Silvex	-Atrazine
-Aldicarb	-Phthalates
-Chlordane	-Acrylamide
-Dalapon	-Dibromochloropropane
-Diquat	

Table A.4 (Continued)

Synthetic Organics continued:

-Endothall	-1,2-Dichloropropane
-Epichlorohydrin	-Pentachlorophenol
-Glyphosate	-Dibromomethane*
-Toluene	-Pichloram
-Carbofuran	-Xylene
-Adipates	-Dinoseb
-Alachlor	-Hexachlorocyclopentadiene
-Dioxin	-Ethylene dibromide

Radionuclides:

-Radium 226 and 228	-Uranium
-Beta particle and photon radioactivity	-Gross alpha particle activity
	-Radon

* EPA may make substitutions for up to 7 listed compounds. The following were proposed for removal from this list in July 1987 for the listed reasons below:

Zinc and Silver: No adverse health effects at levels found in drinking water.

Aluminum and Sodium: Unresolved issues on adverse effects.

Dibromomethane, Molybdenum, and Vanadium: Insufficient health effects data.

At the same time candidates for removal were proposed, the following candidates for addition to this list were proposed: Aldicarb, Sulfoxide, Aldicarb sulfone, Ethylbenzene, Heptachlor, Heptachlor expoxide, Styrene, and Nitrite.

Table A.5
DOT Hazard Classifications

Classification	Hazard definition	Examples
Explosive	A chemical mixture or device designed to function by explosion	Bombs
Explosive A	Detonating or otherwise of maximum hazard (Explosive A can detonate)	Grenades, dynamite, detonating fusees, small-arms ammunition
Explosive B	Flammable hazard (can burn violently or propulsively)	Propellant explosives, airplane flare incendiary grenades
Explosive C	Minimum hazard (can burn briskly)	Explosive rivets, igniter cord, common fireworks
Blasting	A material designed for blasting, having little probability of accidental explosion	Ammonium nitrate-fuel oil mixture
Flammable Liquid	Liquid having flash point below 100 degrees fahrenheit	Gasoline, ethyl alcohol, leather dressing
Combustible Liquid	Liquid having flash point at or above 100 degrees fahrenheit	Kerosene, alcohol solution (24%) fuel oil
Flammable Solid	Any solid other than explosive which is liable to cause fire through friction	Fusees, phosphorous, camera film, lithium metal, oily rags

Table A.5 (Continued)

Classification	**Hazard Definition**	**Examples**
Oxidizer	A substance that yields oxygen readily to stimulate combustion (can sensitize other materials to easy ignition or rapid burning)	Barium nitrate, potassium permanganate aluminum nitrate
Organic Peroxide	An organic derivative of hydrogen peroxide - bleaches by oxidation (can react or burn violently)	Acetyl peroxide solution
Corrosive Material	A liquid or solid that causes visible destruction of human skin tissue or severe corrosion of metal	Sulfuric acid, wet storage batteries
Flammable Gas	A gas flammable under certain pressure and temperature (gas under pressure in a container that if released will burn readily)	Acetylene, dimethyl, ether
Nonflammable Gas	A gas which is not flammable or poisonous (not likely to burn if released, but but can harm people on exposure)	Carbon dioxide, fluorine, argon
Poison A	A poison gas or liquid a very small amount of which is dangerous to life (can be deadly in small amounts)	Insecticide, liquefied gas, poison grenade
Poison B	A poisonous liquid or solid so toxic to humans as to cause a hazard to health (takes more and works slower, but can injure seriously)	Parathion, liquid disinfectant, endrin

Classification	Hazard Definition	Examples
Irritating Material	A liquid or solid which gives off dangerous or intensely irritating fumes when exposed to fire or air	Tear gas, chemical ammunition
Etiologic Agent	A viable microorganism which may cause human disease	Toxin in blood and liquid samples shipped for diagnosis, biological product used for experimentation
Radioactive Material	Fissile radioactive material	Radioactive devices, uranium metal
Other Regulated Materials (ORM)	Any material not meeting the above hazard class definitions but which may pose an unreasonable risk when transported. These materials are divided into five classes:	
ORM-A	A material which can cause extreme annoyance or discomfort to people	Aldrin, acetylene tetrabromide
ORM-B	A material capable of causing significant damage to a vehicle from leakage	Ferrous chloride, gallium metal, bone oil
ORM-C	A material with properties which make it unsuitable for shipment	Garbage tankage, castor beans, dry ice
ORM-D	A consumer commodity which presents a limited hazard	Charcoal briquets, propane cigarette lighters, aerosol spray

Table A.5 (Continued)

Classification	Hazard Definition	Examples
ORM-E	A material not in any other hazard class	Hazardous waste, hazardous mixture
Forbidden	Materials which are prohibited from being offered or accepted for transportation	Methylene glycol dinitrate, mercury nitride, loose mixture of sulfur and chlorate

Table A.6
Hazardous Wastes from Specific Sources

Industry/EPA Hazardous Waste No.	Hazardous Waste
Wood Preservation:	
K001	Bottom sediment sludge from the treatment of wastewaters from wood preserving processes that use creosote and/or pentachlorophenol
Inorganic Pigments:	
K002	Wastewater treatment sludge from the production of chrome yellow and orange pigments
K003	Wastewater treatment sludge from the production of molybdate orange pigments
K004	Wastewater treatment sludge from the production of zinc yellow pigments
K005	Wastewater treatment sludge from the production of chrome green pigments
K006	Wastewater treatment sludge from the production of chrome oxide green pigments (anhydrous and hydrated)
K007	Wastewater treatment sludge from the production of iron blue pigments
K008	Oven residue from the production of chrome oxide green pigments
Organic Chemicals:	
K009	Distillation bottoms from the production of acetaldehyde from ethylene
K010	Distillation side cuts from the production of acetaldehyde from ethylene
K011	Bottom stream from the wastewater stripper in the production of acrylonitrile

Table A.6 (Continued)

Industry/EPA Hazardous Waste No.	Hazardous Waste
K013	Bottom stream from the acetonitrile column in the production of acrylonitrile
K014	Bottoms from the acetonitrile purification column in the production of acrylonitrile
K015	Still bottoms from the distillation of benzyl chloride
K016	Heavy ends or distillation residues from the production of carbon tetrachloride
K017	Heavy ends (still bottoms) from the purification column in the production of epichlorohydrin
K018	Heavy ends from the fractionation column in ethyl chloride production
K019	Heavy ends from the distillation of ethylene dichloride in ethylene dichloride production
K020	Heavy ends from the distillation of vinyl chloride in vinyl chloride monomer production
K021	Aqueous spent antimony catalyst waste from fluoromethanes production
K022	Distillation bottom tars from the production of phenol/acetone from cumene
K023	Distillation light ends from the production of phthalic anhydride from napthalene
K024	Distillation bottoms from the production of phthalic anhydride from napthalene
K093	Distillation light ends from the production of phthalic anhydride from ortho-xylene
K094	Distillation bottoms from the production of phthalic anhydride from ortho-xylene

Industry/EPA Hazardous Waste No.	Hazardous Waste
K025	Distillation bottoms from the production of nitrobenzene by the nitration of benzene
K026	Stripping still tails from the production of methyl ethyl pyridines
K027	Centrifuge and distillation residues from toluene diisocyanate production
K028	Spent catalyst from the hydrochlorinator reactor in the production of 1,1,1-trichloroethane
K029	Waste from the product steam stripper in the production of 1,1,1-trichloroethane
K095	Distillation bottoms from the production of 1,1,1-trichloroethane
K096	Heavy ends from the heavy ends column from the production of 1,1,1-trichloroethane
K030	Column bottoms or heavy ends from the combined production of trichloroethylene and perchloroethylene
K083	Distillation bottoms from aniline production
K103	Process residues from aniline extraction from the production of aniline
K104	Combined wastewater streams generated from nitrobenzene/aniline production
K085	Distillation or fractionation column bottoms from the production of chlorobenzenes
K105	Separate aqueous stream from the reactor product washing step in the production of chlorobenzenes
K111	Product washwater from the production of dinitrotoluene via nitration of toluene

Table A.6 (Continued)

Industry/EPA Hazardous Waste No.	Hazardous Waste
K112	Reaction by-product water from the drying column in the production of toluenediamine via hydrogenation of dinitrotoluene
K113	Condensed liquid light ends from the purification of toluenediamine in the production of toluenediamine via hydrogenation of dinitrotoluene
K114	Vicinals from the purification of toluenediamine in the production of toluenediamine via hydrogenation of dinitrotoluene
K115	Heavy ends from the purification of toluenediamine in the production of toluenediamine via hydrogenation of dinitrotoluene
K116	Organic condensate from the solvent recovery column in the production of toluene diisocyanate via phosgenation of toluenediamine
K117	Wastewater from the reactor vent gas scrubber in the production of ethylene dibromide via bromination of ethene
K118	Spent adsorbent solids from purification of ethylene dibromide in the production of ethylene dibromide via bromination of ethene
K136	Still bottoms from the purification of ethylene dibromide in the production of ethylene dibromide via bromination of ethene
Inorganic Chemicals:	
K071	Brine purification muds from the mercury coil process in chlorine production, where separately prepurified brine is not used
K073	Chlorinated hydrocarbon waste from the purification step of the diaphragm cell process using graphite anodes in chlorine production

Industry/EPA Hazardous Waste No.	Hazardous Waste
K106	Wastewater treatment sludge from the mercury coil process in chlorine production
Pesticides:	
K031	By-product salts generated in the production of MSMA and cacodylic acid
K032	Wastewater treatment sludge from the production of chlordane
K033	Wastewater and scrub water from the chlorination of cyclopentadiene in the production of chlordane
K034	Filter solids from the filtration of hexachlorocyclopentadiene in the production of chlordane
K097	Vacuum stripper discharge from the chlordane chlorinator in the production of chlordane
K035	Wastewater treatment sludges generated in the production of creosote
K036	Still bottoms from toluene reclamation distillation in the production of disulfoton
K037	Wastewater treatment sludges from the production of disulfoton
K038	Wastewater from the washing and stripping of phorate production
K039	Filter cake from the filtration of diethylphosphorodithioic acid in the production of phorate
K040	Wastewater treatment sludge from the production of phorate
K041	Wastewater treatment sludge from the production of toxaphene

Table A.6 (Continued)

Industry/EPA Hazardous Waste No.	Hazardous Waste
K098	Untreated process wastewater from the production of toxaphene
K042	Heavy ends or distillation residues from the distillation of tetrachlorobenzene in the production of 2,4,5-T
K043	2,6-Dichlorophenol waste from the production of 2,4-D
K099	Untreated wastewater from the production of 2,4-D
Explosives Industry:	
K044	Wastewater treatment sludges from the manufacturing and processing of explosives
K045	Spent carbon from the treatment of wastewater containing explosives
K046	Wastewater treatment sludges from the manufacturing, formulation and loading of lead-based initiating compounds
K047	Pink/red water from TNT operations
Petroleum Refining Industry:	
K048	Dissolved air flotation (DAF) float from the petroleum refining industry
K049	Slop oil emulsion solids from the petroleum refining industry
K050	Heat exchanger bundle cleaning sludge from the petroleum refining industry
K051	API separator sludge from the petroleum refining industry
K052	Tank bottoms (leaded) from the petroleum refining industry

Industry/EPA Hazardous Waste No.	Hazardous Waste
Iron and Steel Industry:	
K061	Emission control dust/sludge from the primary production of steel in electric furnaces
K062	Spent pickle liquor generated by steel finishing operation of plants that produce iron or steel
Secondary Lead Industry:	
K069	Emission control dust/sludge from secondary lead smelting
K100	Waste leaching solution from acid leaching of emission control dust/sludge from secondary lead smelting
Veterinary Pharmaceutical Industry:	
K084	Wastewater treatment sludges generated during the production of veterinary pharmaceuticals from arsenic or organo-arsenic compounds
K101	Distillation tar residues from the distillation of aniline-based compounds in the production of veterinary pharmaceuticals from arsenic or organo-arsenic compounds
K102	Residue from the use of activated carbon for decolorization in the production of veterinary pharmaceuticals from arsenic or organo-arsenic compounds
Ink Formulation Industry:	
K086	Solvent washes and sludges, caustic washes and sludges, or water washes and sludges from cleaning tubs and equipment used in the formulation of ink from pigments, driers, soaps, and stabilizers containing chromium and lead

Table A.6 (Continued)

Industry/EPA Hazardous Waste No.	Hazardous Waste
Coking Industry:	
K060	Ammonia still lime sludge from coking operations
K087	Decanter tank tar sludge from coking operations

Table A.7
Acute Hazardous Wastes

No.	Substance
P023	Acetaldehyde, chloro-
P002	Acetamide, N-(amino-thioxomethyl)-
P057	Acetamide, 2-fluoro-
P058	Acetic acid, fluoro-, sodium salt
P066	Acetimidic acid, N-] (methylcar- bamoyl) oxy]thio-, methyl ester
P001	3-(alpha-Acetonyl-benzyl)-4-hydroxy-coumarin and salts, when present at con-centrations greater than 0.3%
P002	1-Acetyl-2-thiourea
P003	Acrolein
P070	Aldicarb
P004	Aldrin
P005	Allyl alcohol
P006	Aluminum phosphide
P007	5-(Aminomethyl)-3-isoxazolol
P008	4-aAminopyridine
P009	Ammonnium picrate (R)
P119	Ammonium vanadate
P010	Arsenic acid
P012	Arsenic (III) oxide
P011	Arsenic (V) oxide
P011	Arsenic pentoxide
P012	Arsenic trioxide
P038	Arsine, diethyl-
P054	Aziridine
P013	Barium cyanide
P024	Benzenamine, 4-chloro-
P077	Benzenamine, 4-nitro-
P028	Benzene,(chloro-methyl)-
P042	1,2-Benzenediol,4-[1-hydroxy-2-methyl amino)ethyl]-
P014	Benzenethiol
P028	Benzyl chloride
P015	Beryllium dust
P016	Bis(chloromethyl) ether
P017	Bromoacetone
P018	Brucine
P021	Calcium cyanide
P123	Camphene, octachloro-
P103	Carbamimidoselenoic acid
P022	Carbon bisulfide
P022	Carbon disulfide
P095	Carbonyl chloride
P033	Chlorine cyanide
P023	Chloroacetaldehyde
P024	p-Chloroaniline
P026	1-(o-Chlorophenyl) thiourea
P027	3-Chloropropionitrile
P029	Copper cyanides
P030	Cyanides (soluble cyanide salts), not elsewhere specified
P031	Cyanogen
P033	Cyanogen chloride
P036	Dichlorophenylarsine
P037	Dieldrin
P038	Diethylarsine
P039	O,O-Diethyl S-[2-(ethylthio)ethyl] phosphorodithioate
P041	Diethyl-p-nitrophenyl phosphate
P040	0,0-Diethyl O-pyrazinyl phosphorothioate
P043	Diisopropyl fluoro-phosphate
P044	Dimethoate
P045	3,3-Dimethyl-1-(methyl-thio)-2-butanone, O-[(methylamino) carbonyl] oxime

Table A.7 (Continued)

No.	Substance	No.	Substance
P071	O,O-Dimethyl O-p-nitrophenyl phosphorothioate	**P037**	1,2,3,4,10,10-Hexachloro-6,7-epoxy-1,4,4a,5,6,7,8,8a-octahydro-endo,exo-1,4:5,8 demethanonaphthalene
P082	Dimethylnitrososamine	**P060**	1,2,3,4,10,10-Hexachloro -1,4,4a,5,8,8a-hexahydro-1,4:5,8-endo, endo-dimeth- anonaphthalene
P046	alpha, alpha-Dimethylphenethylamine	**P004**	1,2,3,4,10,10-Hexachloro-1,4,4a,5,8,8a-hexahydro-1,4:5,8-endo, exodimethanonaphthalene
P047	4,6-Dinitro-o-cresol and salts	**P060**	Hexachlorohexahydro-exo, exodimethanonaphthalene
P034	4,6-Dinitro-o-cyclohexylphenol	**P062**	Hexaethyl tetraphosphate
P084	2,4-Dinitrophenol	**P116**	Hydrazinecarbothioamide
P020	Dinoseb	**P068**	Hydrazine, methyl-
P085	Diphosphoramide, octamethyl-	**P063**	Hydrocyanic acid
P039	Disulfoton	**P063**	Hydrocyanic cyanide
P049	2,4-Dithiobiuret	**P096**	Hydrogen phosphide
P109	Dithlopyrophosphoric acid, tetraethyl ester	**P064**	Isocyanic acid, methyl ester
P050	Endosulfan	**P007**	3(2H)-Isoxazolone, 5-(aminomethyl)-
P088	Endothall	**P092**	Mercury, (acetato-O) phenyl-
P051	Endrin	**P065**	Mercury fulminate (R,T)
P042	Epinephrine	**P016**	Methane, oxybis(chloro-
P046	Ethanamine, 1,1-dimethyl-2-phenyl-	**P112**	Methane, tetranitro-(R)
P084	Ethenamine, N-methyl-N-nitroso-	**P118**	Methanethiol, trichloro-
P101	Ethyl cyanide	**P059**	4,7-Methano-1H-indene, 1,4,5,6,7,8,8-heptachloro-3a,4,7,7a-tetrahydro-
P054	Ethylenimine	**P066**	Methomyl
P097	Famphur	**P067**	2-Methylaziridine
P056	Fluorine	**P068**	Methyl hydrazine
P057	Fluoroacetamide	**P064**	Methyl isocyanate
P058	Fluoroacetic acid, sodium salt	**P069**	2-Methyllactonitrile
P065	Fulminic acid, mercury(II) salt (R,T)	**P071**	Methyl parathion
P059	Heptachlor	**P072**	alpha-Naphthylthiourea
P051	1,2,3,4,10,10-Hexachloro-6,7-epoxy-1,4,4a,5,6,7,8,8a-octahydro-endo, endo-1,4:5,8-dimethanonaphthalene	**P073**	Nickel carbonyl
		P074	Nickel cyanide

No.	Substance	No.	Substance
P074	Nickel(II) cyanide	**P041**	Phosphoric acid, diethyl p-nitrophenyl ester
P073	Nickel tetracarbonyl	**P044**	Phosphorodithioic acid, O,O-dimethyl S-[2-(methylamino)-2-oxoethyl]ester
P075	Nicotine and salts	**P043**	Phosphorofluoric acid, bis(1-methylethyl)-ester
P076	Nitric oxide	**P094**	Phosphorothioic acid, O,O-diethyl S-(ethylthio)methyl ester
P077	p-Nitroaniline	**P089**	Phosphorothioic acid, O,O-diethyl O-(p-nitrophenyl) ester
P078	Nitrogen dioxide	**P040**	Phosphorothioic acid, O,O-diethyl O-pyrazinyl ester
P076	Nitrogen(II) oxide	**P097**	Phosphorothioic acid, O,O-dimethyl O-[p-((dimethylamino)-sulfonyl)phenyl]ester
P078	Nitrogen(IV) oxide	**P110**	Plumbane, tetraethyl-
P081	Nitroglycerine (R)	**P098**	Potassium cyanide
P082	N-Nitrosodimethyl-amine	**P099**	Potassium silver cyanide
P084	N-Nitrosomethyl-vinylamine	**P070**	Propanal, 2-methyl-2-(methylthio)-, O-[(methylamino) carbonyl] oxime
P050	5-Norbornene-2,3-dimethanol, 1,4,5, 6,7,7-hexachloro, cylic sulfite	**P101**	Propanenitrile
P085	Octamethylpyrophosph-oramide	**P027**	Propanenitrile, 3-chloro-
P087	Osmium oxide	**P069**	Propanenitrile, 2-hydroxy-2-methyl-
P087	Osmium tetroxide	**P081**	1,2,3-Propanetriol, trinitrate-(R)
P088	7-Oxabicyclo[2.2.1] heptane-2,3-dicarboxylic acid	**P017**	2-Propanone, 1-bromo-
P089	Parathion	**P102**	Propargyl alcohol
P034	Phenol, 2-cyclohexyl-4,6-dinitro-		
P048	Phenol,2,4-dinitro-		
P047	Phenol,2,4-dinitro-6-methyl-		
P020	Phenol,2,4-dinitro-6-(1-methylpropyl)-		
P009	Phenol,2,4,6-trinitro-, ammonium salt (R)		
P036	Phenyl dichloroarsine		
P092	Phenylmercuric acetate		
P093	N-Phenylthiourea		
P094	Phorate		
P095	Phosgene		
P096	Phosphine		

Table A.7 (Continued)

No.	Substance	No.	Substance
P003	2-Propenal	**P093**	Thiourea, phenyl-
P005	2-Propen-1-ol	**P123**	Toxaphene
P067	1,2-Propylenimine	**P118**	Trichloromethanethiol
P102	2-Propyn-1-ol	**P119**	Vanadic acid, ammonium salt
P008	4-Pyridinamine	**P120**	Vanadium pentoxide
P075	Pyridine, (S)-3-(methyl-2-pyrrolidinyl)-, and salts	**P120**	Vanadium(V)oxide
P111	Pyrophosphoric acid, tetraethyl ester	**P001**	Warfarin, when present at concentrations greater than 0.3%
P103	Selenourea	**P121**	Zinc cyanide
P104	Silver cyanide	**P122**	Zinc phosphide (R,T)
P105	Sodium azide	**P122**	Zinc phosphide, when present at concentrations greater than 10%
P106	Sodium cyanide		
P107	Strontium sulfide		
P108	Strychnidin-10-one, and salts		
P018	Strychnidin-10-one, 2,3-dimethoxy-		
P108	Strychnine and salts		
P115	Sulfuric acid, thallium(I) salt		
P109	Tetraethyldithiopyrophosphate		
P110	Tetraethyl lead		
P111	Tetraethylpyrophosphate		
P112	Tetranitromethane (R)		
P062	Tetraphosphoric acid, hexaethyl ester		
P113	Thallic oxide		
P113	Thallium(III) oxide		
P114	Thallium(I) selenite		
P115	Thallium(I) sulfate		
P045	Thiofanox		
P049	Thioimidodicarbonic diamide		
P014	Thiophenol		
P116	Thiosemicarbazide		
P026	Thiourea, (2-chlorophenyl)-		
P072	Thiourea, 1-naphthalenyl-		

Table A.8
Hazardous Wastes from Nonspecific Sources

Industry/EPA Hazardous Waste No.	Hazardous Waste
Generic:	
F001	The following spent halogenated solvents used in degreasing: Tetrachloroethylene, trichloroethylene, methylene chloride, 1,1,1-trichloroethane, carbon tetrachloride, and chlorinated fluorocarbons; all spent solvent mixtures/blends used in degreasing containing, before use, a total of ten percent or more (by volume) of one or more of the above halogenated solvents or those solvents listed in F002, F004, and F005; and still bottoms from the recovery of these spent solvents and spent solvent mixtures.
F002	The following spent halogenated solvents: Tetrachloroethylene, methylene chloride, trichloroethylene, 1,1,1-trichloroethane, clorobenzene, 1,1,2-trichloro-1,2,2-trifluoroethane, ortho-dichlorobenzene, trichlorofluoromethane, and 1,1,2-trichloroethane; all spent solvent mixtures/blends containing, before use, a total of ten percent or more (by volume) of one or more of the above halogenated solvents or those listed in F001, F004, or F005; and still bottoms from the recovery of these spent solvents and spent solvent mixtures.
F003	The following spent non-halogenated solvents: Xylene, acetone, ethyl acetate, ethylbenzene, ethyl ether, methyl isobutyl ketone, n-butyl alcohol, cyclohexanone, and methanol; all spent solvent mixtures/blends containing, before use, only the above spent non-halogenated solvents; and all spent solvent mixtures/blends containing before use, one or more of the above non-halogenated solvents, and a total of ten percent or more (by volume) of one or more of those solvents listed in F001, F002, F004, and F005; and still bottoms from the recovery of these spent solvents and spent solvent mixtures.

Table A.8 (Continued)

Industry/EPA Hazardous Waste No.	Hazardous Waste
F004	The following spent non-halogenated solvents: Cresols and cresylic acid and nitrobenzene; all spent solvent mixtures/blends containing, before use, a total of ten percent or more (by volume) of one or more of the above non-halogenated solvents or those solvents listed in F001, F002, F005; and still bottoms from the recovery of these spent solvents and spent solvent mixtures.
F005	The following spent non-halogenated solvents: Toluene, methyl ethyl ketone, carbon disulfide, isobutanol, pyridine, benzene, 2-ethoxyethanol, and 2-nitropropane; all spent solvent mixtures/ blends containing, before use, a total of ten percent or more (by volume) of one or more of the above non-halogenated solvents or those solvents listed in F001, F002, or F004; and still bottoms from the recovery of these spent solvents and spent solvent mixtures.
F006	Wastewater treatment sludges from electroplating operations except from the following processes: (1) sulfuric acid anodizing of aluminum; (2) tin plating on carbon steel; (3) zinc plating (segregated basis) on carbon steel; (4) aluminum or zinc-aluminum plating on carbon steel; (5) cleaning/stripping associated with tin, zinc and aluminum plating on carbon steel; and (6) chemical etching and milling of aluminum.
F019	Wastewater treatment sludges from the chemical conversion coating of aluminum...
F007	Spent cyanide plating bath solutions from electroplating operations...
F008	Plating bath residues from the bottom of plating baths from electroplating operations where cyanides are used in the process.

Industry/EPA Hazardous Waste No.	Hazardous Waste
F009	Spent stripping and cleaning bath solutions from electroplating operations where cyanides are used in the process.
F010	Quenching bath residues from oil baths from metal heat treating operations where cyanides are used in the process.
F011	Spent cyanide solutions from salt bath pot cleaning from metal heat treating operations.
F012	Quenching waste water treatment sludges from metal heat treating operations where cyanides are used in the process.
F024	Wastes, including but not limited to, distillation residues, heavy ends, tars, and reactor clean-out wastes from the production of chlorinated aliphatic hydrocarbons, having carbon content from one to five, utilizing free radical catalyzed processes. (This listing does not include light ends, spent filters and filter aids, spent dessicants, wastewater, wastewater treatment sludges, spent catalysts, and wastes listed in 261.32.)
F020	Wastes (except wastewater and spent carbon from hydrogen chloride purification) from the production or manufacturing use (as a reactant, chemical intermediate, or component in a formulating process) of tri- or tetrachlorophenol, or of intermediates used to produce their pesticide derivatives. (This listing does not include wastes from the production of Hexachlorophene from highly purified 2,4,5-trichlorophenol.)
F021	Wastes (except wastewater and spent carbon from hydrogen chloride purification from the production or manufacturing use (as a reactant, chemical intermediate, or component in a formulating process) of pentachlorophenol, or of intermediates used to produce its derivatives.

Table A.8 (Continued)

Industry/EPA Hazardous Waste No.	Hazardous Waste
F022	Wastes (except wastewater and spent carbon from hydrogen chloride purification) from the manufacturing use (as a reactant, chemical intermediate, or component in a formulating process) of tetra-, penta-, or hexachlorobenzenes under alkaline conditions.
F023	Wastes (except wastewater and spent carbon from hydrogen chloride purification) from the production of materials on equipment previously used for the production or manufacturing use (as a reactant, chemical intermediate, or component in a formulating process) of tri- and tetrachlorophenols. (This listing does not include wastes from equipment used only for the production or use of Hexachlorophene from highly purified 2,4,5-trichlorophenol.)
F026	Wastes (except wastewater and spent carbon from hydrogen chloride purification) from the production of materials on equipment previously used for the manufacturing use (as a reactant, chemical intermediate, or component in a formulating process) of tetra-, penta-, or hexachlorobenzene under alkaline conditions.
F027	Discarded unused formulations containing tri-, tetra, or pentachlorophenol or discarded unused formulations containing compounds derived from those chlorophenols (this listing does not include formulations containing Hexachlorophene synthesized from prepurified 2,4,5-trichlorophenol as the sole component).
F028	Residues resulting from the incineration or thermal treatment of soil contaminated with EPA Hazardous Waste Nos. F020, F021, F022, F023, F026, and F027.

Table A.9
Toxic Hazardous Wastes

No.	Substance	No.	Substance
U001	Acetaldehyde (I)	**U010**	carbonyl) oxy)
U034	Acetaldehyde, trichloro-	**(cont)**	methyl]-1,1a,2,8,8a, 8b-hexahydro-8a-methoxy-5-methyl-
U187	Acetamide, N-(4-othoxyphenyl)-	**U157**	Benz[j]aceanthrylene, 1,2-dihydro-3-methyl-
U005	Acetamide, N-9H-fluoren-2-yl-		
U112	Acetic acid, ethyl ester (I)	**U016**	Benz[c]acridine
		U016	3,4-Benzacridine
U144	Acetic acid, lead salt	**U017**	Benzal chloride
		U018	Benz[a]anthracene
U214	Acetic acid, thallium (I) salt	**U018**	1,2-Benzanthracene
		U094	1,2-Benzanthracene, 7,12-dimethyl-
U002	Acetone (I)		
U003	Acetonitrile (I,T)	**U012**	Benzenamine (I,T)
U248	3-(alpha-Acetonyl-benzyl)-4-hydroxy-courmarin and salts when present at concentrations of 0.3% or less	**U014**	Benzenamine, 4,4´-carbonimidoylbis (N,N- di-methyl-
		U049	Benzenamine, 4-chloro-2-methyl
		U093	Benzenamine, N,N´-dimethyl-4-phenylazo-
U004	Acetophenone		
U005	2-Acetylaminofluorene		
U006	Acetyl chloride (C,R,T)	**U158**	Benzenamine, 4,4´-methylenebis(2-chloro-
U007	Acrylamide		
U008	Acrylic acid (I)	**U222**	Benzenamine, 2-,methyl-, hydrochloride
U009	Acrylonitrile		
U150	Alanine, 3-[p-bis(2-chloroethyl)amino] phenyl-, L-	**U181**	Benzenamine, 2-methyl-5-nitro
		U019	Benzene (I,T)
U328	2-Amino-I-methyl-benzene	**U038**	Benzeneacetic acid, 4-chloro-alpha-(4-chlorophenyl)-alpha-hydroxy, ethyl ester
U353	4-Amino-I-methyl-benzene		
U011	Amitrole	**U030**	Benzene, 1-bromo-4-phenoxy-
U012	Aniline (I,T)		
U014	Auramine	**U037**	Benzne, chloro-
U015	Azaserine	**U190**	1,2-Benzenedicarbo-xylic acid anhydride
U010	Azirino(2´,3´:3,4) pyrrolo(1,2-a) indole-4,7-dione, 6-amino-8-[((amino-	**U028**	1,2-Benzenedicarbo-xylic acid, [bis(2-ethyl-hexyl)]ester

Table A.9 (Continued)

No.	Substance	No.	Substance
U069	1,2-Benzenedicarboxylic acid, dibutyl ester	U055	Benzene, (1-methylethyl)-(I)
U088	1,2-Benzenedicarboxylic acid, diethyl ester	U169	Benzene, nitro-(I,T)
U102	1,2-Benzenedicarboxylic acid, dimethyl ester	U183	Benzene, pentachloro-
U107	1,2-Benzenedicarboxylic acid, di-n-octyl ester	U185	Benzene, pentachloronitro-
U070	Benzene, 1,2-dichloro-	U020	Benzenesulfonic acid chloride (C,R)
U071	Benzene, 1,3-dichloro-	U020	Benzenesulfonyl chloride (C,R)
U072	Benzene, 1,4-dichloro-	U207	Benzene, 1,2,4,5-tetrachloro-
U017	Benzene,(dichloromethyl)-	U023	Benzene, (trichloromethyl)-(C,R,T)
U223	Benzene, 1,3-diisocyanatomethyl-(R,T)	0234	Benzene, 1,3,5-trinitro- (R,T)
U239	Benzene, dimethyl-(I,T)	U021	Benzidine
U201	1,3-Benzenediol	U202	1,2-Benzisothiazolin-3-one, 1,1-dioxide
U127	Benzene, hexachloro-	U120	Benzo[j,k]fluorine
U056	Benzene, hexahydro-(I)	U022	Benzo[a]pyrene
U188	Benzene, hydroxy-	U022	3,4-Benzopyrene
U220	Benzene, methyl-	U197	p-Benzoquinone
U105	Benzene, 1-methyl-1-2,4-dinitro-	U023	Benzotrichloride (C,R,T)
U106	Benzene, 1-methyl-2,6-dinitro-	U050	1,2-Benzphenanthrene
U203	Benzene, 1,2-methylenedioxy-4-allyl-	U085	2,2´-Bioxirane (I,T)
U141	Benzene, 1,2-methylenedioxy-4-propenyl-	U021	(1,1´-Biphenyl)-4,4´-diamine
U090	Benzene, 1,2-methylenedioxy-4-propyl-	U073	(1,1´-Biphenyl)-4,4´-diamine, 3,3´-dichloro-
		U091	(1,1´-Biphenyl)-4,4´-diamine, 3,3´-dimethoxy
		U095	(1,1´-Biphenyl)-4,4´-diamine, 3,3´-dimethyl-
		U024	Bis(2-chloroethoxy) methane
		U027	Bis(2-chloroisopropyl) ether
		U244	Bis(dimethylthio,-carbamoyl) disulfide

No.	Substance
U028	Bis(2-ethylhexyl) phthalate
U246	Bromine cyanide
U225	Bromoform
U030	4-Bromophenyl phenyl ether
U128	1,3-Butadiene, 1,1,2, 3,4,4-hexachloro-
U172	1-Butanamine, N-butyl-N-nitroso-
U035	Butanoic acid, 4-[Bis (2-chloroethyl) amino] benzene-
U031	1-Butanol (I)
U159	2-Butanone (I,T)
U160	2-Butanone peroxide (R,T)
U053	2-Butenal
U074	2-Butene, 1,4-dichloro- (I,T)
U031	n-Butyl alcohol (I)
U136	Cacodylic acid
U032	Calcium chromate
U238	Carbamic acid, ethyl ester
U178	Carbamic acid, methylnitroso-, ethyl ester
U176	Carbamide, N-ethyl-N-nitroso-
U177	Carbamide, N-methyl-N-nitroso-
U219	Carbamide, thio-
U097	Carbamoyl chloride, dimethyl-
U215	Carbonic acid, dithallium(I) salt
U156	Carbonochloridic acid, methyl ester (I,T)
U033	Carbon oxyfluoride (R,T)
U211	Carbon tetrachloride
U033	Carbonyl fluoride (R,T)
U034	Chloral
U035	Chlorambucil
U036	Chlordane, technical
U026	Chlornaphazine
U037	Chlorobenzene
U039	4-Chloro-m-cresol
U041	1-Chloro-2,3-epoxy-propane
U042	2-Chloroethyl vinyl ether
U044	Chloroform
U046	Chloromethyl methyl ether
U047	beta-Chloronaph-thalene
U048	o-Chlorophenol
U049	4-Chloro-o-toluidine, hydrochloride
U032	Chromic acid, calcium salt
U050	Chrysene
U051	Creosote
U052	Cresols
U052	Cresylic acid
U053	Crotonaldehyde
U055	Cumene (I)
U246	Cyanogen bromide
U197	1,4-Cyclohexadiene-dione
U056	Cyclohexane (I)
U057	Cyclohexanone (I)
U130	1,3-Cyclopentadiene, 1,2,3,4,5,5-hexa-chloro-
U058	Cyclophosphamide
U240	2,44-D, salts and esters
U059	Daunomycin
U060	DDD
U061	DDT

Table A.9 (Continued)

No.	Substance	No.	Substance
U142	Decachlorooctahydro-1,3,4-metheno-2H-cyclobuta[c,d]-pentalen-2-one	U085	1,2:3,4-Diepoxybutane (I,T)
U062	Diallate	U108	1,4-Diethylene dioxide
U133	Diamine (R,T)	U086	N,N-Diethylhydrazine
U221	Diaminotoluene	U087	O,O-Diethyl-S-methyl-dithiophosphate
U063	Dibenz[a,h]anthracene	U088	Diethyl phthalate
U063	1,2:5,6-Dibenzanthracene	U089	Diethylstilbestrol
U064	1,2:7,8-Dibenzopyrene	U148	1,2-Dihydro-3,6-pyradizinedione
U064	Dibenz[a,i]pyrene	U090	Dihydrosafrole
U066	1,2-Dibromo-3-chloropropane	U091	3,3´Dimethoxybenzidine
U069	Dibutyl Phthalate	U092	Dimethylamine (I)
U062	S-(2,3-Dichloroallyl) diisopropylthiocarbamate	U093	Dimethylaminoazobenzene
U070	o-Dichlorobenzene	U094	7,12-Dimethylbenz[a]-anthracene
U071	m-Dichlorobenzene	U095	3,3´-Dimethylbenzidine
U072	p-Dichlorobenzene	U096	alpha,alpha-Dimethylbenzylhydroperoxide (R)
U073	3,3´-Dichlorobenzidine	U097	Dimethylcarbamoyl chloride
U074	1,4-Dichloro-2-butene (I,T)	U098	1,1-Dimethylhydrazine
U075	Dichlorodifluoromethane	U099	1,2-Dimethylhydrazine
U192	3,5-Dichloro-N-(1,1-dimethyl-2-propynyl) benzamide	U101	2,4-Dimethylphenol
U060	Dichloro diphenyl dichloroethane	U102	Dimethyl phthalate
U061	Dichloro diphenyl trichloroethane	U103	Dimethyl sulfate
U078	1,1-Dichloroethylene	U105	2,4-Dinitrotoluene
U079	1,2-Dichloroethylene	U106	2,6-Dinitrotoluene
U025	Dichloroethyl ether	U107	Di-n-octyl phthalate
U081	2,4-Dichlorophenol	U108	1,4-Dioxane
U082	2,6-Dichlorophenol	U109	1,2-Diphenylhydrazine
U240	2,4-Dichlorophenoxyacetic acid, salts and esters	U110	Dipropylamine (I)
U083	1,2-Dichloropropane	U111	Di-N-propylnitrosamine
U084	1,3-Dichloropropene	U001	Ethanal (I)
		U174	Ethanamine, N-ethyl-N-nitroso-
		U067	Ethane, 1,2-dibromo-
		U076	Ethane, 1,1-dichloro-
		U077	Ethane, 1,2-dichloro-

No.	Substance	No.	Substance
U114	1,2-Ethanodiylbis-carbamodithioic acid	**U114**	Ethylenebis(dithio-carbamic acid)
U131	Ethane, 1,1,1,2,2,2-hexachloro-	**U067**	Etylene dibromide
		U077	Ethylene dichloride
U024	Ethane, 1,1´-[methyl-enebis(oxy)]bis[2-chloro-	**U115**	Ethlene oxide (I,T)
		U116	Ethylene thiourea
		U117	Ethyl ether (I)
U003	Ethanenitrile (I,T)	**U076**	Ethylidene dichloride
U117	Ethane, 1,1´-oxybis-(I)	**U118**	Ethylmethacrylate
		U119	Ethyl methanesulfo-nate
U025	Ethane, 1,1´-oxybis [2-chloro-	**U139**	Ferric dextran
U184	Ethane, pentachloro-	**U120**	Fluoranthene
U208	Ethane, 1,1,1,2-tetrachloro-	**U122**	Formaldehyde
		U123	Formic acid (C,T)
U209	Ethane, 1,1,2,2-tetrachloro-	**U124**	Furan (I)
		U125	2-Furancarboxaldehyde (I)
U218	Ethanethioamide		
U247	Ethane, 1,1,1,-trichloro-2,2-bis(p-methoxyphenyl),	**U147**	2,5-Furandione
		U213	Furan, tetrahydro-(I)
U227	Ethane, 1,1,2-trichloro-	**U125**	Furfural (I)
		U124	Furfuran (I)
U043	Ethene, chloro-	**U206**	D-Glucopyranose, 2-deoxy-2(3-methyl-3-nitro-soureldo)-
U042	Ethene, 2-chloroeth-oxy-		
U078	Ethene, 1,1-dichloro-	**U126**	Glycidylaldehyde
U079	Ethene, trans-1,2-dichloro-	**U163**	Guanidine, N-nitroso-N-methyl-N´nitro-
U210	Ethene, 1,1,2,2-tetrachloro-	**U127**	Hexachlorobenzene
		U128	Hexachlorobutadiene
U173	Ethanol, 2,2´-(nitro-solmino)bis-	**U129**	Hexachlorocyclohexane (gamma isomer)
U004	Ethanone, 1-phenyl-	**U130**	Hexachlorocyclopen-tadiene
U006	Ethanoyl chloride (C,R,T)		
		U131	Hexachloroethane
U359	2-Ethoxyethanol.	**U132**	Hexachlorophene
U112	Ethyl acetate (I)	**U243**	Hexachloropropene
U113	Ethyl acrylate (I)	**U133**	Hydrazine (R,T)
U238	Ethyl carbamate (urethan)	**U086**	Hydrazine, 1,2-diethyl-
U038	Ethyl 4,4´-dichloro-benzilate	**U098**	Hydrazine, 1,1-dimethyl-
U359	Ethylene glycol mono-ethyl ether.	**U099**	Hydrazine, 1,2-dimethyl-

Table A.9 (Continued)

No.	Substance	No.	Substance
U109	Hydrazine,1,2-diphenyl	**U119**	Methanesulfonic acid, ethyl ester
U134	Hydrofluoric acid (C,T)	**U211**	Methane, tetrachloro-
U134	Hydrogen fluoride (C,T)	**U121**	Methane, trichloro-fluoro-
U135	Hydrogen sulfide	**U153**	Methanethiol (I,T)
U096	Hydroperoxide, 1-methyl-1-phenyl-ethyl- (R)	**U225**	Methane, tribromo-
		U044	Methane, trichloro
		U121	Methane, trichloro-fluoro-
U136	Hydroxydimethylarsine oxide	**U123**	Methanoic acid (C,T)
U116	2-Imidazolidnethione	**U036**	4,7-Methanoindan, 1,2,4,5,6,7,8,8-octachloro-3a,4,7,7a-tetrahydro-
U137	Indeno[1,2,3-cd]-pyrene		
U139	Iron dextran		
U140	Isobutyl alcohol (I,T)	**U154**	Methanol (I)
		U155	Methapyrilene
U141	Isosafrole	**U247**	Methoxychlor.
U142	Kepone	**U154**	Methyl alcohol (I)
U143	Laslocarpine	**U029**	Methyl bromide
U144	Lead acetate	**U186**	1-Methylbutadiene
U145	Lead phosphate	**U045**	Methyl chloride (I,T)
U146	Lead subacetate	**U156**	Methyl chlorocarbonate (I,T)
U129	Lindane		
U147	Maleic anhydride	**U226**	Methylchloroform
U148	Maleic hydrazide	**U157**	3-Methylcholanthrene
U149	Malononitrile	**U158**	4,4´-Methylenebis(2-chloroaniline)
U150	Melphalan		
U151	Mercury	**U132**	2,2´-Methylenebis (3,4,6,-trichloro-phenol)
U152	Methacrylonitrile (I,T)		
U092	Methanamine, N-methyl-(I)	**U068**	Methylene bromide
		U080	Methylene chloride
U029	Methane, bromo-	**U122**	Methylene oxide
U045	Methane, chloro- (I,T)	**U159**	Methyl ethyl ketone (I,T)
U046	Methane, chloromethoxy	**U160**	Methyl ethyl ketone peroxide (R,T)
U068	Methane, dibromo-	**U138**	Methyl Iodide
U080	Methane, dichloro	**U161**	Methyl isobutyl ketone (I)
U075	Methane, dichlorodi-fluoro-		
U138	Methane, iodo-	**U162**	Methyl methacrylate (I,T)

No.	Substance	No.	Substance
U163	N-Methyl-N´-nitro-N-nitrosoguanidine	**U177**	N-Nitroso-N-methyl-urea
U161	4-Methyl-2-pentanone (I)	**U178**	N-Nitroso-N-methylur-ethane
U164	Methylthiouracil	**U180**	N-Nitrosopyrroli-dine
U010	Mitomycin C	**U181**	5-Nitro-o-toluidine
U059	5,12-Naphthacenedione, (8S-cis)-8-acetyl-10-[3-amino-2,3,6-trideoxy- alpha-L-lyxo-hexopyranosyl) oxyl]- 7,8,9,10-tetrahydro- 6,8,11-tri- hydroxy-1-methoxy-	**U193**	1,2-Oxathiolane, 2,2-dioxide
		U058	2H-1,3,2-Oxazaphos-phorine, 2-[bis(2-chloroethyl)amino] tetrahydro-,oxide 2-
		U115	Oxirane (I,T)
		U041	Oxirane, 2-(chloro-methyl)-
U165	Naphthelene	**U182**	Paraldehyde
U047	Naphthalane, 2-chloro-	**U183**	Pentachlorobenzene
U166	1,4-Naphthalenedione	**U184**	Pentachloroethane
U236	2,7-Naphthalenedisul-fonic acid, 3,3-[3,3-dimethyl-(1,1´-biphenyl)-4,4´diyl)]-bis (azo)bis(5-amino-4- hydroxy)-tetra-sodium salt	**U185**	Pentachloronitro-benzene
		See F027	Pentachlorophenol
		U186	1,3-Pentadiene (I)
		U187	Phenacetin
U166	1,4, Naphthaquinone	**U188**	Phenol
U167	1-Naphthylamine	**U048**	Phenol, 2-chloro-
U168	2-Naphthylamine	**U039**	Phenol, 4-chloro-3-methyl-
U167	alpha-Naphthylamine		
U168	beta-Naphthylamine	**U081**	Phenol, 2,4-dichloro
U026	2-Naphthylamine, N,N´-bis(2-chloromethyl)-	**U082**	Phenol, 2,6-dichloro
		U101	Phenol, 2,4-dimethyl-
U169	Nitrobenzene (I,T)	**U170**	Phenol, 4-nitro-
U170	p-Nitrophenol	**SEE F027**	Phenol, pentachloro- Do Phenol, 2,3,4,6-tetrachloro- Do Phenol, 2,4,5-trich-loro- Do Phenol, 2,4,6-trichloro-
U171	2-Nitropropane (I,T)		
U172	N-Nitrosodi-n-butyl-amine		
U173	N-Nitrosodiethanol-amine		
U174	N-Nitrosodiethylamine		
U111	N-Nitroso-N-propyl-amine	**U137**	1,10-(1,2-phenylene) pyrene
U176	N-Nitroso-N-ethylurea	**U145**	Phosphoric acid, Lead salt

Table A.9 (Continued)

No.	Substance
U087	Phosphorodithioic acid, 0,0-diethyl-, S-methylester
U189	Phosphorous sulfide (R)
U190	Phthalic anhydride
U179	N-Nitrosopiperidine
U191	2-Picoline
U192	Pronamide
U194	1-Propanamine (I,T)
U110	1-Propanamine, N-propyl-(I)
U066	Propane, 1,2-dibromo-3-chloro-
U149	Propanedinitrile
U171	Propane, 2-nitro-(I,T)
U017	Propane, 2,2´oxybis [2-chloro-
U193	1,3-Propane sultone
U235	1-Propanol, 2,3-dibromo-,phosphate (3:1)
U126	1-Propanol, 2,3-epoxy
U140	1-Propanol, 2-methyl-(I,T)
U002	2-Propanone (I)
U007	2-Propenamide
U084	Propene, 1,3-dichloro-
U243	1-Propene, 1,1,2,3,3,3-hexachloro-
U009	2-Propenenitrile
U152	2-Propenenitrile, 2-methyl-(I,T)
U008	2-Propenoic acid (I)
U113	2-Propenoic acid, ethlyl ester (I)
U118	2-Propenoic acid, 2-methyl-, ethyl ester
U162	2-Propenoic acid, 2-methyl-,ethyl ester (I,T)
SEE F027	Proplonic acid, 2-(2,4,5-trichlorophenoxy)-
U194	n-Propylamine (I,T)
U083	Propylene dichloride
U196	Pyridine
U155	Pyridine, 2-[(2-(dimethlamino)-2-thenylamino]-
U179	Pyridine, hexahydro-N-nitroso-
U191	Pyridine, 2-methyl-
U164	4(1H)-Pyrimidinone, 2,3-dihydro-6-methyl-2-thioxo-
U180	Pyrrole, tetrahydro-N-nitroso-
U200	Reserpine
U201	Resorcinol
U202	Saccharin and salts
U203	Safrole
U204	Selenious acid
U204	Selenium dioxide
U205	Selenium disulfide (R,T)
U015	L-Serine, diazoacetate (ester)
SEE F027	Silvex
U089	4,4´-Stilbenediol, alpha,alpha´-diethyl-
U206	Streptozotocin
U135	Sulfur hydride
U103	Sulfuric acid, dimethyl ester
U189	Sulfur phosphide (R)
U205	Sulfur selenide (R,T)
SEE F027	2,4,5-T
U207	1,2,4,5-Tetrachlorobenzene

No.	Substance	No.	Substance
U208	1,1,1,2-Tetrachloro-ethane	**U121**	Trichloromonofluoro-methane
U209	1,1,2,2-Tetrachloro-ethane	**SEE F027**	2,4,5-Trichlorophenol
U210	Tetrachloroethylene	**Do**	2,4,6-Trichlorophenol
SEE F027	2,3,4,6-Tetrachloro-phenol	**Do**	2,4,5-Trichloro-phenoxy-acetic acid
U213	Tetrahydrofuran (I)	**U234**	sym-Trinitrobenzene (R,T)
U214	Thallium(I) acetate	**U182**	1,3,5-Trioxane, 2,4,5-trimethyl-
U215	Thallium(I) carbonate	**U235**	Tris(2,3-dibromo-propyl) phosphate
U216	Thallium(I) chloride	**U236**	Trypan blue
U217	Thallium(I) nitrate	**U237**	Uracil 5[bis(2-chloro-methyl)amino]-
U218	Thioacetamide	**U237**	Uracil mustard
U153	Thiomethanol (I,T)	**U043**	Vinyl Chloride
U219	Thiourea	**U248**	Warfarin, when present at concentrations of 0.3% or less
U244	Thiram	**U239**	Xylene (I)
U220	Toluene	**U200**	Yohimban-16-carbo-xylic acid, 11,17-dimethoxy-18-[(3,4,5-trimethoxy-benzoyl) oxy]-, methyl ester
U221	Toluenediamine	**U249**	Zinc phosphide, when present at concen-trations of 10% or less
U223	Toluene diisocyanate (R,T)		
U328	o-Toluidine		
U222	O-Toluidine hydro-chloride		
U353	p-Toluidine		
U011	1H-1,2,4-Triazol-3-amine		
U226	1,1,1-Trichloroethane		
U227	1,1,2-Trichloroethane		
U228	Trichloroethane		
U228	Trichloroethylene		

The primary hazardous properties of these materials have been indicated by the letters T (Toxicity), R (Reactivity), I (Ignitability), and C (Corrosivity). Absence of a letter indicates that the compound is only listed for toxicity.

Table A.10
National Priorities List: Non-Federal Sites

NPL Rank	ST	Site Name	Township/City Borough/County
1	NJ	Lipari Landfill	Pitman
2	DE	Tybouts Corner Landfill	New Castle
3	PA	Bruin Lagoon	Bruin
4	NJ	Helen Kramer Landfill	Mantua
5	MA	Industri-Plex	Woburn
6	NJ	Price Landfill	Pleasantville
7	NY	Pollution Abatement Services	Oswego
8	IA	LaBounty Site	Charles City
9	DE	Army Creek Landfill	New Castle
10	NJ	CPS/Madison Industries	Old Bridge
11	MA	Nyanza Chemical Waste Dump	Ashland
12	NJ	GEMS Landfill	Gloucester
13	MI	Berlin & Farro	Swartz Creek
14	MA	Baird & McGuire	Holbrook
15	NJ	Lone Pine Landfill	Freehold
16	NH	Somersworth Sanitary Landfill	Somersworth
17	MN	FMC Corp. (Fridley Plant)	Fridley
18	AR	Vertac, Inc.	Jacksonville
19	NH	Keefe Environmental Services	Epping
20	MT	Silver Bow Creek/Butte Area	Sil Bow/Deer Lg
21	SD	Whitewood Creek	Whitewood
22	TX	French, Ltd.	Crosby
23	NH	Sylvester	Nashua
24	MI	Liquid Disposal, Inc.	Utica
25	PA	Tysons Dump	Upper Merion
26	PA	McAdoo Associates	McAdoo
27	TX	Motco Inc.	La Marque
28	OH	Arcanum Iron & Metal	Darke
29	MT	East Helena Site	East Helena
30	TX	Sikes Disposal Pits	Crosby
31	AL	Triana/Tennessee River	Limestn./Morgan
32	CA	Stringfellow	Glen Avon Hgts.
33	ME	McKin Co.	Gray
34	TX	Crystal Chemical Co.	Houston
35	NJ	Bridgeport Rental & Oil Ser.	Bridgeport
36	CO	Sand Creek Industrial	Commerce
37	TX	Geneva Indust./Fuhrmann Energy	Houston
38	MA	W. R. Grace & Co. (Acton Plant)	Acton
39	MN	New Brighton/Arden Hills	New Brighton
40	FL	Skhuylkill Metals Corp.	Plant City
41	NJ	Vineland Chemical Co., Inc.	Vineland
42	NJ	Burnt Fly Bog	Marlboro

NPL Rank	ST	Site Name	Township/City Borough/County
43	MN	Reilly Tar (St. Louis Park Plnt.)	St. Louis Park
44	NY	Old Bethpage Landfill	Oyster Bay
45	FL	Reeves SE Galvanizing Corp.	Tampa
46	NJ	Shieldalloy Corp.	Newfield
47	MT	Anaconda Co. Smelter	Anaconda
48	WA	Western Processing Co., Inc.	Kent
49	WI	Omega Hills North Landfill	Germantown
50	FL	American Creosote (Pensacola Pit)	Pensacola
51	NJ	Caldwell Trucking Co.	Fairfield
52	NY	GE Moreau	S. Glen Falls
53	FL	Peak Oil Co./Bay Drum Co.	Tampa
54	OH	United Scrap Lead Co., Inc.	Troy
55	OK	Tar Creek (Ottawa County)	Ottawa
56	KS	Cherokee County	Cherokee
57	IN	Seymour Recycling Corp.	Seymour
58	NJ	Brick Township Landfill	Brick
59	MI	Northernaire Plating	Cadillac
60	WA	Frontier Hard Chrome, Inc.	Vancouver
61	WI	Janesville Old Landfill	Janesville
62	SC	Independent Nail Co.	Beaufort
63	SC	Kalama Specialty Chemicals	Beaufort
64	WI	Janesville Ash Beds	Janesville
65	FL	Davie Landfill	Davie
66	OH	Miami County Incinerator	Troy
67	FL	Gold Coast Oil Corp.	Miami
68	IN	International Minerals (E. Plt.)	Terre Haute
69	WI	Wheeler Pit	La Prairie
70	AZ	Tucson Intl Airport Area	Tucson
71	CA	Operating Indust., Inc. Lndfll	Monterey Park
72	NY	Wide Beach Development	Brant
73	CA	Iron Mountain Mine	Redding
74	NJ	Scientific Chemical Processing	Carlstadt
75	CO	California Gulch	Leadville
76	NJ	D´Imperio Property	Hamilton
77	MN	Oakdale Dump	Oakdale
78	MI	Gratiot County Landfill	St. Louis
79	RI	Picillo Farm	Coventry
80	MA	New Bedford Site	New Bedford
81	LA	Old Inger Oil Refinery	Darrow
82	OH	Chem-Dyne	Hamilton
83	SC	SCRDI Bluff Road	Columbia
84	CT	Laurel Park, Inc.	Naugatuck

Table A.10 (Continued)

NPL Rank	ST	Site Name	Township/City Borough/County
85	CO	Marshall Landfill	Boulder
86	IL	Outboard Marine Corp.	Waukegan
87	NM	South Valley	Albuquerque
88	VT	Pine Street Canal	Burlington
89	WV	West Virginia Ordnance	Point Pleasant
90	MO	Ellisville Site	Ellisville
91	ND	Arsenic Trioxide Site	Southeastern ND
92	VA	Matthews Electroplating	Roanoke
93	IA	Aidex Corp.	Council Bluffs
94	AZ	Mountain View Mobile Home	Globe
95	TN	North Hollywood Dump	Memphis
96	KY	A.L. Taylor (Valley of Drums)	Brooks
97	GU	Ordot Landfill	Guam
98	MS	Flowood Site	Flowood
99	UT	Rose Park Sludge Pit	Salt Lake City
100	KS	Arkansas City Dump	Arkansas City
101	IL	Parsons Casket Hardware Co.	Belvidere
102	IL	A & F Material Reclaiming, Inc.	Greenup
103	PA	Douglassville Disposal	Douglassville
104	NJ	Krysowaty Farm	Hillsborough
105	MN	Koppers Coke	St. Paul
106	MA	Plymouth Harbor/Cannon Eng.	Plymouth
107	ID	Bunker Hill Mining & Metallurg	Smelterville
108	NY	Hudson River PCBs	Hudson River
109	NJ	Universal Oil Prod. (Chem Div)	East Rutherford
110	CA	Aerojet General Corp.	Rancho Cordova
111	WA	Com Bay, South Tacoma Channel	Tacoma
112	PA	Osborne Landfill	Grove City
113	UT	Portland Cmnt. (Kiln Dust 2 & 3)	Salt Lake City
114	CT	Old Southington Landfill	Southington
115	NY	Syosset Landfill	Oyster Bay
116	AZ	Nineteenth Avenue Landfill	Phoenix
117	OR	Teledyne Wah Chang	Albany
118	WA	Midway Landfill	Kent
119	NY	Sinclair Refinery	Wellsville
120	AL	Mowbray Engineering Co.	Greenville
121	MI	Spiegelberg Landfill	Green Oak
122	FL	Miami Drum Services	Miami
123	NJ	Reich Farms	Pleasant Plains
124	ID	Union Pacific Railroad Co.	Pocatello
125	NJ	South Brunswick Landfill	South Brunswick
126	AL	Ciba-Geigy Corp. (McIntosh Plt.)	McIntosh
127	FL	Kassauf-Kimerling Battery	Tampa

NPL Rank	ST	Site Name	Township/City Borough/County
128	IL	Wauconda Sand & Gravel	Wauconda
129	TX	Bailey Waste Disposal	Bridge City
130	NH	Ottati & Goss/Kingston Stl. Drum	Kingston
131	MI	Ott/Story/Cordova	Dalton
132	MI	Thermo-Chem, Inc.	Muskegon
133	VA	Greenwood Chemical Co.	Newtown
134	NJ	NL Industries	Pedricktown
135	MN	St. Regis Paper Co.	Cass Lake
136	NJ	Ringwood Mines/Landfill	Ringwood
137	FL	Whitehouse Oil Pits	Whitehouse
138	GA	Hercules 009 Landfill	Brunswick
139	NY	Jones Sanitation	Hyde Park
140	MI	Velsicol Chemical (Michigan)	St. Louis
141	OH	Summit National	Deerfield
142	NY	Love Canal	Niagra Falls
143	DE	Coker´s Sanitation Service Lfs	Kent
144	MI	Rockwell International (Allegan)	Allegan
145	MN	Pine Bend Sanitary Landfill	Dakota
146	IA	Lawrence Todtz Farm	Camanche
147	IN	Fisher-Calo	LaPorte
148	FL	Pioneer Sand Co.	Warrington
149	MI	Springfield Township Dump	Davisburg
150	PA	Hranica Landfill	Buffalo
151	NC	Martin Marietta, Sodyeco, Inc.	Charlotte
152	FL	Zellwood Ground Water Contamin.	Zellwood
153	MI	Packaging Corp. of America	Filer City
154	WI	Muskego Sanitary Landfill	Muskego
155	NY	Hooker (S Area)	Niagara Falls
156	PA	Lindane Dump	Harrison
157	CO	Central City-Clear Creek	Idaho Springs
158	NJ	Ventron/Velsicol	Wood Ridge
159	FL	Taylor Road Landfill	Seffner
160	RI	Western Sand & Gravel	Burrillville
161	SC	Koppers Co., Inc. (Florence Plt.)	Florence
162	NJ	Maywood Chemical Co.	Mywd./Rchel.Pk.
163	NJ	Nascolite Corp.	Millville
164	OH	Industrial Excess Landfill	Uniontown
165	OK	Hardage/Criner	Criner
166	MI	Rose Township Dump	Rose
167	MN	Waste Disposal Engineering	Andover
168	NY	Liberty Industrial Finishing	Farmingdale
169	NJ	Kin-Buc Landfill	Edison
170	IN	Waste Inc. Landfill	Michigan City

Table A.10 (Continued)

NPL Rank	ST	Site Name	Township/City Borough/County
171	OH	Bowers Landfill	Circleville
172	NJ	Ciba-Geigy Corp.	Toms River
173	MI	Butterworth #2 Landfill	Grand Rapids
174	NJ	American Cyanamid Co.	Bound Brook
175	PA	Heleva Landfill	North Whitehall
176	NJ	Ewan Property	Shamong
177	NY	Bativia Landfill	Bativia
178	MN	Boise Cascade/Onan/Medtronics	Fridley
179	RI	L&RR, Inc.	N. Smithfield
180	PA	Butler Mine Tunnel	Pittston
181	FL	NW 58th Street Landfill	Hialeah
182	NJ	Delilah Road	Egg Harbor
183	PA	Mill Creek Dump	Erie
184	NJ	Glen Ridge Radium Site	Glen Ridge
185	NJ	Montclair/W. Orange Radium Site	Mtclr./W. Orng.
186	FL	Sixty-Second Street Dump	Tampa
187	MI	G&H Landfill	Utica
188	NC	Celanese (Shelby Fiber Oper.)	Shelby
189	NJ	Metaltec/Aerosystems	Franklin
190	WI	Schmalz Dump	Harrison
191	MI	Motor Wheel, Inc.	Lansing
192	NJ	Lang Property	Pemberton
193	TX	Stewco, Inc.	Waskom
194	NJ	Sharkey Landfill	Prsipny./Try Hl
195	CA	Selma Treating Co.	Selma
196	LA	Cleve Reber	Sorrento
197	IL	Velsicol Chemical (Illinois)	Marshall
198	MI	Tar Lake	Mancelona
199	NY	Johnstown City Landfill	Town Johnstown
200	NC	NC State U (L. 86, Farm Unit #1)	Raleigh
201	CO	Lowry Landfill	Arapahoe
202	MN	MacGillis & Gibbs/Bell Lumber	New Brighton
203	PA	Hunterstown Road	Straban
204	MD	Woodlawn County Landfill	Woodlawn
205	NJ	Combe Fill North Landfill	Mount Olive
206	MA	Re-Solve, Inc.	Dartmouth
207	NJ	Goose Farm	Plumstead
208	TN	Velsicol Chem (Hardeman County)	Toone
209	NY	York Oil Co.	Moira
210	FL	Sapp Battery Salvage	Cottondale
211	SC	Wamchem, Inc.	Burton
212	NJ	Chemical Leaman Tank Lines, Inc.	Bridgeport
213	WI	Master Disposal Service Lndfil	Brookfield

NPL Rank	ST	Site Name	Township/City Borough/County
214	KS	Doepke Disposal (Holliday)	Johnson
215	NJ	Florence Land Recontour. Lndfll	Florence
216	RI	Davis Liquid Waste	Smithfield
217	MA	Charles-George Reclam. Lndfll	Tyngsborough
218	NJ	King of Prussia	Winslow
219	VA	Chisman Creek	York
220	OH	Nease Chemical	Salem
221	CO	Eagle Mine	Minturn/Rdclif.
222	NJ	W. R. Grace & Co. (Wayne Plant)	Wayne
223	NJ	Chemical Control	Elizabeth
224	NC	Charles Macon Lagoon & Drum Stor.	Rock Hill
225	SC	Leonard Chemical Co., Inc.	Rock Hill
226	OH	Allied Chemical & Ironton Coke	Ironton
227	MI	Verona Well Field	Battle Creek
228	MO	Lee Chemical	Liberty
229	CT	Beacon Heights Landfill	Beacon Falls
230	AL	Stauffer Chem (Cold Creek Plant)	Bucks
231	MN	Burlington Northern (Brainerd)	Brainerd/Baxter
232	MI	Torch Lake	Houghton
233	RI	Central Landfill	Johnston
234	PA	Malvern TCE	Malvern
235	NY	Facet Enterprises, Inc.	Elmira
236	DE	Delaware Sand & Gravel Landfill	New Castle
237	PA	MW Manufacturing	Valley
238	VA	C & R Battery Co., Inc.	Chesterfield
239	TN	Murray-Ohio Dump	Lawrenceburg
240	IN	Envirochem Corp.	Zionsville
241	IN	MIDCO I	Gary
242	OH	Ormet Corp.	Hannibal
243	OH	South Point Plant	South Point
244	PA	Whitmoyer Laboratories	Jackson
245	FL	Coleman-Evans Wood Preserv. Co.	Whitehouse
246	NJ	Dayco Corp./L.E. Carpenter Co.	Wharton
247	PA	Shriver´s Corner	Straban
248	PA	Dorney Road Landfill	Upper Macungie
249	IN	Northside Sanitary Lndfl, Inc.	Zionsville
250	FL	Florida Steel Corp.	Indiantown
251	IL	Pagel´s Pit	Rockford
252	MN	U of Minnesota Rosemnt Res Cent	Rosemount
253	MN	Freeway Sanitary Landfill	Burnsville
254	AZ	Litchfield Airport Area	Goodyear/Avndl.
255	CA	Firestone Tire (Salinas Plant)	Salinas
256	NJ	Spence Farm	Plumstead

Table A.10 (Continued)

NPL Rank	ST	Site Name	Township/City Borough/County
257	AR	Mid-South Wood Products	Mena
258	MS	Newsome Brothers/Old Reichhold	Columbia
259	CA	Atlas Asbestos Mine	Fresno
260	CA	Coalinga Asbestos Mine	Coalinga
261	FL	Brown Wood Preserving	Live Oak
262	NY	Port Washington Landfill	Port Washington
263	IN	Columbus Old Mnincpl. Lndfll #1	Columbus
264	NJ	Combe Fill South Landfill	Chester
265	NJ	JIS Landfill	Jsbrg./S. Bnsk.
266	NY	Tronic Plating Co., Inc.	Farmingdale
267	PA	Centre County Kepone	State College
268	OH	Fields Brook	Ashtabula
269	CT	Solvents Recovery Service	Southington
270	CO	Woodbury Chemical Co.	Commerce City
271	NJ	Waldick Areospace Devices, Inc.	Wall
272	MA	Hocomonco Pond	Westborough
273	KY	Distler Brickyard	West Point
274	NY	Ramapo Landfill	Ramapo
275	CA	Coast Wood Preserving	Ukiah
276	CA	South Bay Asbestos Area	Alviso
277	NY	Mercury Refining, Inc.	Colonie
278	FL	Hollingsworth Solderless Terminal	Fort Lauderdale
279	NY	Olean Well Field	Olean
280	FL	Varsol Spill	Miami
281	MN	Joslyn Manufactg. & Supply Co.	Brooklyn Center
282	PA	York County Solid Waste/Refuse Lf	Hopewell
283	WI	Spickler Landfill	Spencer
284	CO	Denver Radium Site	Denver
285	PA	Route 940 Drum Dump	Pocono Summit
286	FL	Tower Chemical Co.	Clermont
287	PA	C & D Recycling	Foster
288	MO	Syntex Facility	Verona
289	MT	Milltown Reservoir Sediments	Milltown
290	MN	Arrowhead Refinery Co.	Hermantown
291	OR	Martin-Marietta Aluminum Co.	The Dalles
292	CO	Uravan Uranium (Union Carbide)	Uravan
293	NJ	Pijak Farm	Plumstead
294	NJ	Syncon Resins	South Kearney
295	MN	Oak Grove Sanitary Landfill	Oak Grove
296	CA	Liquid Gold Oil Corp.	Richmond
297	CA	Purity Oil Sales, Inc.	Malaga
298	NH	Tinkham Garage	Londonderry
299	FL	Alpha Chemical Corp.	Galloway

NPL Rank	ST	Site Name	Township/City Borough/County
300	NJ	Bog Creek Farm	Howell
301	ME	Saco Tannery Waste Pits	Saco
302	PR	Frontera Creek	Rio Abajo
303	FL	Pickettville Road Landfill	Jacksonville
304	OH	Alsco Anaconda	Gnadenhutten
305	MA	Iron Horse Park	Billerica
306	PA	Palmerton Zinc Pile	Palmerton
307	IN	Neal´s Landfill (Bloomington)	Bloomington
308	WI	Kohler Co. Landfill	Kohler
309	AL	Interstate Lead Co. (ILCO)	Leeds
310	AZ	Hassayampa Landfill	Hassayampa
311	MA	Silresim Chemical Corp.	Lowell
312	MA	Wells G&H	Woburn
313	NJ	Chemsol, Inc.	Piscataway
314	WI	Lauer I Sanitary Landfill	Menomomee Falls
315	MI	Petoskey Municipal Well Field	Petoskey
316	MN	Union Scrap	Minneapolis
317	NJ	Radiation Technology, Inc.	Rockaway
318	NJ	Fair Lawn Well Field	Fair Lawn
319	IN	Main Street Well Field	Elkhart
320	MN	Lehillier/Mankato Site	Lehillier/Mnkt.
321	WA	Lakewood Site	Lakewood
322	PA	Industrial Lane	Williams
323	IN	Fort Wayne Reduction Dump	Fort Wayne
324	WI	Onalaska Municipal Landfill	Onalaska
325	WI	National Presto Indust., Inc.	Eau Claire
326	NJ	Monroe Township Landfill	Monroe
327	NJ	Rockaway Borough Well Field	Rockaway
328	IN	Wayne Waste Oil	Columbia City
329	MD	Mid-Atlantic Wood Prsrvrs., Inc.	Harmans
330	ID	Pacific Hide & Fur Recyc. Co.	Pocatello
331	IA	Des Moines TCE	Des Moines
332	NJ	Beachwood/Berkley Wells	Berkley
333	NY	Vestal Water Supply Well 4-2	Vestal
334	PR	Vega Alta Public Supply Wells	Vega Alta
335	MI	Sturgis Municipal Wells	Sturgis
336	MN	Washington County Landfill	Lake Elmo
337	TX	Odessa Chromium #1	Odessa
338	TX	Odessa Chromium #2 (Andrews Hwy)	Odessa
339	NE	Hastings Ground Water Contamin	Hastings
340	AZ	Indian Bend Wash Area	Scottsdale/
340	AZ	Indian Bend Wash Area	Tmpe./ Phoenix
341	CA	San Gabriel Valley (Area 1)	El Monte

Table A.10 (Continued)

NPL Rank	ST	Site Name	Township/City Borough/County
342	CA	San Gabriel Valley (Area 2)	Baldwin Park
343	CA	San Fernando Valley (Area 1)	Los Angeles
344	CA	San Fernando Valley (Area 2)	Ls Angls./Glnd.
345	CA	San Fernando Valley (Area 3)	Glendale
346	CA	T.H. Agriculture & Nutrition Co.	Fresno
347	WA	Com Bay, Near Shore/Tide Flats	Pierce
348	IL	LaSalle ELectric Utilities	LaSalle
349	IL	Cross Brothers Pail (Pembroke)	Pembroke
350	NC	Jadco-Hughes Facility	Belmont
351	NJ	Monitor Dev./Intercircuits Inc.	Wall
352	PR	Upjohn Facility	Barceloneta
353	CA	McColl	Fullerton
354	PA	Henderson Road	Upper Merion
355	NY	Hooker Chem./Ruco Polymer Corp.	Hicksville
356	WA	Colbert Landfill	Colbert
357	LA	Petro-Processors	Scotlandville
358	NY	Applied Environmental Services	Glenwood Lndg.
359	PR	Barceloneta Landfill	Florida Afuera
360	NH	Tibbets Road	Barrington
361	MD	Sand, Gravel & Stone	Elkton
362	CT	Revere Textile Prints Corp.	Sterling
363	MI	Spartan Chemical Co.	Wyoming
364	NJ	Roebling Steel Co.	Florence
365	PA	East Mountain Zion	Springettsbury
366	TN	Amnicola Dump	Chattanooga
367	NJ	Vineland State School	Vineland
368	MA	Groveland Wells	Groveland
369	NY	General Motors (Cent Foundry Div)	Massena
370	NH	Mottolo Pig Farm	Raymond
371	SC	SCRDI Dixiana	Cayce
372	MI	Roto-Finish Co., Inc.	Kalamazoo
373	MN	Olmsted County Sanitary Landfill	Oronoco
374	MO	Quality Plating	Sikeston
375	MO	Fulbright Landfill	Springfield
376	PA	Presque Isle	Erie
377	NJ	Williams Property	Swaintown
378	NJ	Renora, Inc.	Edison
379	NJ	Denzer & Schafer X-ray Co.	Bayville
380	NJ	Hercules, Inc. (Gibbstown Plant)	Gibbstown
381	IN	Ninth Avenue Dump	Gary
382	SC	Golden Strip Septic Tank Ser.	Simpsonville
383	WA	Toftdahl Drums	Brush Prairie
384	TX	Texarkana Wood Preserving Co.	Texarkana

NPL Rank	ST	Site Name	Township/City Borough/County
385	AR	Gurley Pit	Edmondson
386	FL	Petroleum Products Corp.	Pembroke Park
387	RI	Peterson/Puritan, Inc.	Lincoln/Cmblnd.
388	MO	Times Beach Site	Times Beach
389	MI	Wash King Laundry	Pleasant Plains
390	MN	Whittaker Corp.	Minneapolis
391	WI	Algoma Municipal Landfill	Algoma
392	MN	NL Industries/Taracorp/Golden	St. Louis Park
393	CA	Westinghouse (Sunnyvale Plant)	Sunnyvale
394	CT	Kellog-Deering Well Field	Norwalk
395	MA	Cannon Engineering Corp. (CEC)	Bridgewater
396	MI	H. Brown Co., Inc.	Grand Rapids
397	NY	Nepera Chemical Co., Inc.	Maybrook
398	NY	Niagara County Refuse	Wheatfield
399	FL	Sherwood Medical Industries	Deland
400	AL	Olin Corp. (McIntosh Plant)	McIntosh
401	MI	Southwest Ottawa County Lndfl	Park
402	NY	Kentucky Avenue Well Field	Horseheads
403	NY	Pasley Solvents & Chem., Inc	Hempstead
404	NJ	Asbestos Dump	Millington
405	KY	Lee´s Lane Landfill	Louisville
406	AR	Frit Industries	Walnut Ridge
407	OH	Fultz Landfill	Jackson
408	FL	Tri-City Oil Consvtionist., Inc.	Tampa
409	OH	Coshocton Landfill	Franklin
410	TN	Arlington Blending & Packaging	Arlington
411	RI	Davis (GSR) Landfill	Glocester
412	PA	Lord-Shope Landfill	Girard
413	WA	FMC Corp. (Yakima Pit)	Yakima
414	WI	Northern Engraving Co.	Sparta
415	TX	South Cavalcade Street	Houston
416	MA	PSC Resources	Palmer
417	MI	Forest Waste Products	Otisville
418	PA	Drake Chemical	Lock Haven
419	NH	Kearsarge Metallurgical Corp.	Conway
420	SC	Palmetto Wood Preserving	Dixianna
421	IL	Petersen Sand & Gravel	Libertyville
422	MI	Clare Water Supply	Clare
423	PA	Havertown PCP	Haverford
424	DE	New Castle Spill	New Castle
425	MT	Idaho Pole Co.	Bozeman
426	DE	NCR Corp. (Millsboro Plant)	Millsboro
427	IN	Lake Sandy Jo (M&M Landfill)	Gary

Table A.10 (Continued)

NPL Rank	ST	Site Name	Township/City Borough/County
428	IL	Johns-Manville Corp.	Waukegan
429	MI	Chem Central	Wyoming
430	MI	Novaco Industries	Temperance
431	MN	Windom Dump	Windom
432	NJ	Jackson Township Landfill	Jackson
433	IL	NL Industries/Taracorp Lead Smelt	Granite City
434	MI	K&L Avenue Landfill	Oshtemo
435	WA	Kaiser Aluminum Mead Works	Mead
436	MN	Perham Arsenic Site	Perham
437	MI	Charlevoix Municipal Well	Charlevoix
438	NJ	Montgomery Township Housing Dev	Montgomery
438	NJ	Rocky Hill Municipal Well	Rocky Hill
440	NJ	Cinnaminson Gnd Water Contamin	Cinnaminson
441	NY	Brewster Well Field	Putnam
442	NY	Vestal Water Supply Well 1-1	Vestal
443	PA	Bally Ground Water Contamin	Bally
444	NC	Bypass 601 Ground Water Contamin	Concord
445	MO	Solid State Circuits, Inc.	Republic
446	NE	Waverly Ground Water Contamin	Waverly
447	CA	Advanced Micro Devices, Inc.	Sunnyvale
448	MN	Nutting Truck & Caster Co.	Faribault
449	NJ	U.S. Radium Corp.	Orange
450	TX	Highlands Acid Pit	Highlands
451	PA	Resin Disposal	Jefferson
452	MT	Libby Ground Water Contamination	Libby
453	KY	Newport Dump	Newport
454	PA	Moyers Landfill	Eagleville
455	FL	Parramore Surplus	Mount Pleasant
456	NH	Savage Municipal Water Supply	Milford
457	MN	LeGrand Sanitary Landfill	LaGrand
458	IN	Poer Farm	Hancock
459	PA	Brown´s Battery Breaking	Shoemakersville
460	NY	SMS Instruments, Inc.	Deer Park
461	MI	Hedblum Industries	Oscoda
462	TX	United Creosoting Co.	Conroe
463	NY	Byron Barrel & Drum	Byron
464	WY	Baxter/Union Pacific Tie Treating	Laramie
465	NY	Anchor Chemicals	Hicksville
466	MI	Waste Management-Mich (Holland)	Holland
467	TX	North Cavalcade Street	Houston
468	NJ	Sayreville Landfill	Sayreville
469	NH	Dover Municipal Landfill	Dover
470	NY	Ludlow Sand & Gravel	Clayville

NPL Rank	ST	Site Name	Township/City Borough/County
471	WI	City Disposal Corp. Landfill	Dunn
472	NJ	Tabernacle Drum Dump	Tabernacle
473	NJ	Cooper Road	Voorhees
474	MO	Minker/Stout/Romaine Creek	Imperial
475	KY	Howe Valley Landfill	Howe Valley
476	CT	Yaworski Waste Lagoon	Canterbury
477	WV	Leetown Pesticide	Leetown
478	FL	Cabot/Koppers	Gainesville
479	NJ	Evor Phillips Leasing	Old Bridge
480	PA	William Dick Lagoons	West Caln
481	PA	Wade (ABM)	Chester
482	PA	Lackawanna Refuse	Old Forge
483	OK	Compass Industries (Avery Drive)	Tulsa
484	NJ	Mannheim Avenue Dump	Galloway
485	IN	Neal´s Dump (Spencer)	Spencer
486	NY	Fulton Terminals	Fulton
487	LA	Dutchtown Treatment Plant	Ascension Prsh.
488	PA	Westinghouse Elevator Corp. Pnt.	Gettysburg
489	NH	Auburn Road Landfill	Londonderry
490	WV	Fike Chemical, Inc.	Nitro
491	MN	General Mills/Henkel Corp.	Minneapolis
492	OH	Laskin/Poplar Oil Co.	Jefferson
493	OH	Old Mill	Rock Creek
494	KS	John´s Sludge Pond	Wichita
495	WI	Stoughton City Landfill	Stoughton
496	CA	Del Norte Pesticide Storage	Crescent City
497	NJ	De Rewal Chemical Co.	Kingwood
498	PA	Middletown Air Field	Middletown
499	NJ	Swope Oil & Chemical Co.	Pennsauken
500	GA	Monsanto Corp. (Augusta Plant)	Augusta
501	NH	S. Municipal Water Supply Well	Peterborough
502	ME	Winthrop Landfill	Winthrop
503	WV	Ordnance Works Disposal Areas	Morgantown
504	AR	Cecil Lindsey	Newport
505	OH	Zaresville Well Field	Zanesville
506	NY	Suffern Village Well Field	Vlg. Suffern
507	NY	Endicott Village Well Field	Vlg. Endicott
508	PA	Aladdin Plating	Scott
509	FL	Harris Corp. (Palm Bay Plant)	Palm Bay
510	MN	Kummer Sanitary Landfill	Bemidji
511	OH	Sanitary Landfill Company (IWD)	Dayton
512	WI	Eau Claire Municipal Well Field	Eau Claire
513	MO	Valley Park TCE	Valley Park

Table A.10 (Continued)

NPL Rank	ST	Site Name	Township/City Borough/County
514	CA	San Fernando Valley (Area 4)	Los Angeles
515	CA	Monolithic Memories	Sunnyvale
516	GA	National Semiconductor Corp.	Santa Clara
517	GA	Powersville Site	Peach
518	MI	Grand Traverse Overall Sup. Co.	Greilickville
519	MI	Metamora Landfill	Metamora
520	MI	Whitehall Municipal Wells	Whitehall
521	DE	Stndrd. Chlorine of Deleware, In.	Delaware City
522	MN	South Andover Site	Andover
523	NJ	Diamond Alkali Co.	Newark
524	VA	Avtex Fibers, Inc.	Front Royal
525	MI	Kentwood Landfill	Kentwood
526	MI	Electrovoice	Buchanan
527	NY	Katonah Municipal Well	Town of Bedford
528	CA	Teledyne Semiconductor	Mountain View
529	PR	Fibers Public Supply Wells	Jobos
530	IN	Marion (Bragg) Dump	Marion
531	OH	Pristine, Inc.	Reading
532	WI	Mid-State Disposal, Inc. Lndfl	Cleveland
533	TN	American Creosote (Jackson Plnt)	Jackson
534	CO	Broderick Wood Products	Denver
535	OH	Buckeye Reclamation	St. Clairsville
536	NY	Preferred Plating Corp.	Farmingdale
537	TX	Bio-Ecology Systems, Inc.	Grand Prairie
538	UT	Monticello Rad Contamin. Props	Monticello
539	NJ	Woodland Route 532 Dump	Woodland
540	IN	American Chemical Service, Inc.	Griffith
541	MA	Salem Acres	Salem
542	NY	Richardson Hill Rd. Lndfll/Pond	Sidney Center
543	VT	Old Springfield Landfill	Springfield
544	NY	Solvent Savers	Lincklaen
545	VA	U.S. Titanium	Piney River
546	IL	Galesburg/Koppers Co.	Galesburg
547	NY	Hooker (Hyde Park)	Niagara Falls
548	MI	SCA Independent Landfill	Muskegon Hgts.
549	CA	MGM Brakes	Cloverdale
550	LA	Bayou Sorrell	Bayou Sorrell
551	MI	Duell & Gardner Landfill	Dalton
552	WA	Mica Landfill	Mica
553	NJ	Ellis Property	Evesham
554	KY	Distler Farm	Jefferson
555	CA	Waste Disposal, Inc.	Santa Fe Spngs.
556	WA	Harbor Island (Lead)	Seattle

NPL Rank	ST	Site Name	Township/City Borough/County
557	WI	Lemberger Transport & Recycling	Franklin
558	OH	E.H. Schilling Landfill	Hamilton
559	MI	Cliff/Dow Dump	Marquette
560	NY	Clothier Disposal	Town of Granby
561	PA	Ambler Asbestos Piles	Ambler
562	WA	Queen City Farms	Maple Valley
563	NJ	Curcio Scrap Metal, Inc.	Saddle Brook
564	VA	L.A. Clarke & Son	Spotsylvania
565	WI	Scrap Processing Co., Inc.	Medford
566	MD	Southern Maryland Wood Treating	Hollywood
567	NM	Homestake Mining Co.	Milan
568	CA	Beckman Instruments (Prtrvil.)	Porterville
569	FL	Dubose Oil Products Co.	Cantonment
570	MI	Mason County Landfill	Pere Marquette
571	MI	Cemetery Dump	Rose Center
572	NJ	Hopkins Farm	Plumstead
573	NC	Cape Fear Wood Preserving	Fayetteville
574	RI	Stamina Mills, Inc.	N. Smithfield
575	WI	Lemberger Landfill, Inc.	Whitelaw
576	IN	Reilly Tar (Indianapolis Plant)	Indianapolis
577	ME	Pinette´s Salvage Yard	Washburn
578	TX	Harris (Farley Street)	Houston
579	NJ	Wilson Farm	Plumstead
580	PA	Old City of York Landfill	Seven Valleys
581	PA	Modern Sanitation Landfill	Lower Windsor
582	IL	Byron Salvage Yard	Byron
583	MI	North Bronson Industrial Area	Bronson
584	PA	Stanley Kessler	King of Prussia
585	NJ	Imperial Oil/Champion Chemicals	Morganville
586	NJ	Cosden Chemical Coatings Corp.	Beverly
587	MN	St. Augusta San Lnfl./Engen Dump	St. Augusta
588	NJ	Myers Property	Franklin
589	NJ	Pepe Field	Boonton
590	WA	Northwest Transformer	Everson
591	NY	Genzale Plating Co.	Franklin Square
592	WI	Sheboygan Harbor & River	Sheboygan
593	MI	Ossineke Ground Water Contamin	Ossineke
594	WV	Follansbee Site	Follansbee
595	PA	Keystone Sanitation Landfill	Union
596	NC	Carolina Transformer Co.	Fayetteville
597	NY	North Sea Municipal Landfill	North Sea
598	PA	Bendix Flight Systems Div.	Bridgewater
599	CA	Koppers Co., Inc. (Oroville Pnt.)	Oroville

Table A.10 (Continued)

NPL Rank	ST	Site Name	Township/City Borough/County
600	CA	Louisiana-Pacific Corp.	Oroville
601	MI	S. Macomb Disposal (Lf 9 & 9A)	Macomb
602	MI	U.S. Aviex	Howard
603	PA	Walsh Landfill	Honeybrook
604	NJ	Landfill & Development Co.	Mount Holly
605	NJ	Upper Deerfield Twnsp Sand Lndf	Upper Deerfield
606	NY	Hertel Landfill	Plattekill
607	NY	Haviland Complex	Town Hyde Park
608	NY	Malta Rocket Fuel Area	Malta
609	MI	Kent City Mobile Home Park	Kent City
610	MN	Adrian Municipal Well Field	Adrian
611	NM	AT & SF (Clovis)	Clovis
612	KS	Strother Field Industrial Park	Cowley
613	KS	Obee Road	Hutchinson
614	NJ	Fried Industries	East Brunswick
615	NY	American Thermostat Co.	South Cairo
616	TN	Lewisburg Dump	Lewisburg
617	MI	McGraw Edison Corp.	Albion
618	NY	Goldisc Recordings, Inc.	Holbrook
619	KY	Airco	Calvert City
620	PA	Metal Banks	Philidelphia
621	NY	Sarney Farm	Amenia
622	MA	Rose Disposal Pit	Lanesboro
623	OH	Van Dale Junkyard	Marietta
624	MT	Montana Pole and Treating	Butte
625	KY	B.F. Goodrich	Calvert City
626	MI	Organic Chemicals, Inc.	Grandville
627	NY	Volney Municipal Landfill	Town of Volney
628	NY	FMC Corp. (Dublin Road Lndfl)	Town of Shelby
629	WI	Tomah Fairgrounds	Tomsh
630	MA	Sullivan´s Ledge	New Bedford
631	KY	Smith´s Farms	Brooks
632	PR	Juncos Landfill	Juncos
633	KS	Big River Sand Co.	Wichita
634	IN	Bennett Stone Quarry	Bloomington
635	WA	Wyckoff Co./Eagle Harbor	Bainbridge Isld
636	FL	Munisport Landfill	North Miami
637	AL	Stauffer Chem (LeMoyne Plant)	Axis
638	NJ	M&T Delisa Landfill	Asbury Park
639	TX	Crystal City Airport	Crystal City
640	SC	Geiger (C & M Oil)	Rantoules
641	WI	Moss-American (Kerr-McGee Oil Co)	Milwaukee
642	WI	Waste Research & Reclam. Co.	Eau Claire

NPL Rank	ST	Site Name	Township/City Borough/County
643	OR	Gould, Inc.	Portland
644	NY	Cortese Landfill	Vg. Narrowsburg
645	MN	St. Louis River Site	St. Louis
646	MI	Auto Ion Chemicals, Inc.	Kalamazoo
647	WI	Hagen Farm	Stoughton
648	SC	Carolawn, Inc.	Fort Lawn
649	IA	Midwest Manufacturing/N. Farm	Kellog
650	PA	Berks Sand Pit	Longswamp
651	MI	Sparta Landfill	Sparta
652	IL	ACME Solvent (Morristown Plant)	Morristown
653	NJ	Pomona Oaks Residential Wells	Galloway
654	NY	Rowe Industries Ground Water Cont	Novack/Sag Hrbr
655	PA	Hebelka Auto Salvage Yard	Weisenberg
656	FL	Hipps Road Landfill	Duval
657	MN	Long Prairie Ground Water Contam	Long Prairie
658	MN	Waite Park Wella	Waite Park
659	CA	Applied Materials	Santa Clara
660	CA	Intel Magnetics	Santa Clara
661	CA	Intel Corp. (Santa Clara III)	Santa Clara
662	FL	Pepper Steel & Alloys, Inc.	Medley
663	ME	O´Connor Co.	Augusta
664	WI	Oconomowoc Electrpltng. Co. Inc.	Ashippen
665	MI	Rasmussen´s Dump	Green Oak
666	NY	Kenmark Textile Corp.	Farmingdale
667	PA	Westline Site	Westline
668	KY	Maxey Flats Nuclear Disposal	Hillsboro
669	MT	Mouat Industries	Columbus
670	NY	Claremont Polychemical	Old Bethpage
671	OH	Powell Road Landfill	Dayton
672	PA	Croydon TCE	Croydon
673	IA	Vogel Paint & Wax Co.	Orange City
674	MN	Kurt Manufacturing Co.	Fridley
675	PA	Revere Chemical Co.	Nockamixon
676	MI	Ionia City Landfill	Ionia
677	TX	Koppers Co., Inc. (Texarkana Plt)	Texarkana
678	CO	Lincoln Park	Canon City
679	CO	Smuggler Mountain	Pitkin
680	IN	Wedzeb Enterprises, Inc.	Lebanon
681	PR	GE Wiring Devices	Juana Diaz
682	MI	Ave. "E" Ground Water Contamin	Traverse City
683	OH	New Lyme Landfill	New Lyme
684	NJ	Woodland Route 72 Dump	Woodland
685	PR	RCA Del Caribe	Barceloneta

Table A.10 (Continued)

NPL Rank	ST	Site Name	Township/City Borough/County
686	MN	Koch Refining Co./N-Ren Corp.	Pine Bend
687	PA	Brodhead Creek	Stroudsburg
688	WI	Fadrowski Drum Disposal	Franklin
689	OR	United Chrome Products, Inc.	Corvallis
690	MI	Anderson Development Co.	Adrian
691	WE	Hunts Disposal Landfill	Caledonia
692	MI	Shiawassee River	Howell
693	OK	Tenth Street Dump/Junkyard	Oklahoma City
694	PA	Taylor Borough Dump	Taylor
695	DE	Halby Chemical Co.	New Castle
696	DE	Harvey & Knott Drum, Inc.	Kirkwood
697	TN	Gallaway Pits	Gallaway
698	OH	Big D Campground	Kingsville
699	AR	Midland Products	Ola/Birta
700	NY	Robintech, Inc./National Pipe Co.	Town of Vestal
701	NY	BEC Trucking	Town of Vestal
702	WI	Tomah Armory	Tomah
703	DE	Wildcat Landfill	Dover
704	MI	Burrows Sanitation	Hartford
705	PA	Blosenski Landfill	West Caln
706	VA	Rhinehart Tire Fire Dump	Frederick
707	DE	Delaware City PVC Plant	Delaware City
708	MD	Limestone Road	Cumberland
709	NY	Hooker (102nd Street)	Niagara Falls
710	DE	New Castle Steel	New Castle
711	NM	United Nuclear Corp.	Church Rock
712	PA	Reeser´s Landfill	Upper Macungie
713	AR	Industrial Waste Control	Fort Smith
714	CA	Celtor Chemical Works	Hoopa
715	MA	Haverhill Municipal Landfill	Haverhill
716	AL	Perdido Ground Water Contamin	Perdido
717	NY	Marathon Battery Corp.	Cold Springs
718	NY	Colesville Municipal Landfill	Town Colesville
719	FL	Yellow Water Road Dump	Baldwin
720	OH	Skinner Landfill	West Chester
721	VA	First Piedmont Quarry (Rt. 719)	Pittsylvania
722	NC	Chemtronics, Inc.	Swannanoa
723	IN	MIDCO II	Gary
724	MD	Kane & Lombard Street Drums	Baltimore
725	MO	Shenandoah Stables	Moscow Mills
726	IA	Shaw Avenue Dump	Charles City
727	WA	Silver Mountain Mine	Loomis
728	TX	Petro-Chemical (Turtle Bayou)	Liberty

NPL Rank	ST	Site Name	Township/City Borough/County
729	OH	Republic Steel Corp. Quarry	Elyria
730	MN	Ritari Post & Pole	Sebeka
731	LA	Bayou Bonfouca	Slidell
732	CA	Intel Corp. (Mountain View Plant)	Mountain View
733	CA	Raytheon Corp.	Mountain View
734	MN	Agate Lake Scrapyard	Fairview
735	AR	Jacksonville Municipal Landfill	Jacksonville
736	AR	Rogers Road Municipal Landfill	Jacksonville
737	VA	Saltville Waste Disposal Ponds	Saltville
738	SC	Palmetto Recycling, Inc.	Columbia
739	MA	Shpack Landfill	Nrtn./Attleboro
740	PA	Kimberton Site	Kimberton
741	MA	Norwood PCBs	Norwood
742	MD	Middletown Road Dump	Annapolis
743	WA	Pesticide Lab (Yakima)	Yakima
744	IN	Lemon Lane Landfill	Bloomington
745	IN	Tri-State Plating	Columbus
746	ID	Arrcom (Drexler Enterprises)	Rathdrum
747	NH	Coakley Landfill	North Hampton
748	PA	Fischer & Porter Co.	Warminster
749	CA	Jibboom Junkyard	Sacramento
750	NJ	A. O. Polymer	Sparta
751	WI	Wausau Ground Water Contamin	Wausau
752	NJ	Dover Municipal Well 4	Dover
753	NJ	Rockaway Township Wells	Rockaway
754	WI	Delavan Municipal Well #4	Delavan
755	MO	North-U Drive Well Contamin	Springfield
756	CA	San Gabriel Valley (Area 3)	Alhambre
757	CA	San Gabriel Valley (Area 4)	La Puente
758	WA	American Lake Gardens	Tacoma
759	WA	Greenacres Landfill	Spokane
760	WA	Northside Landfill	Spokane
761	OK	Sand Sprgs Petrochemical Cmplx	Sand Springs
762	TX	Pesses Chemical Co.	Fort Worth
763	MN	East Bethel Demolition Lndfl	East Bethel
764	TX	Triangle Chemical Co.	Bridge City
765	NJ	PJP Landfill	Jersey City
766	PA	Craig Farm Drum	Parker
767	PA	Voortman Farm	Upper Saucon
768	IL	Belvidere Municipal Landfill	Belvidere
769	MO	Bee Cee Manufacturing Co.	Malden
770	PA	Lansdowne Radiation Site	Lansdowne

Table A.11
National Priorities List: Federal Sites

NPL Gr*	ST	Site Name	Township/City Borough/County
2	TN	Milan Army Ammunition Plant	Milan
2	CO	Rocky Mountain Arsenal	Adams
2	CA	McClellan AFB (Grnd Water Cont.)	Sacramento
2	MO	Weldon Sprig Quary (USDOE/Army)	St. Charles
4	GA	Robins AFB (Lndfl #4/Sludge Lag)	Houston
4	NE	Cornhusker Army Ammunition Plant	Hall
4	NJ	Naval Air Engineering Center	Lakehurst
4	UT	Hill Air Force Base	Ogden
6	UT	Odgen Defense Depot	Ogden
6	CA	Sacramento Army Depot	Sacramento
6	IL	Sangamo/Crab Orchard NWR (USDOI)	Carterville
6	ME	Brunswick Naval Air Station	Brunswick
7	WA	McChord AFB (Wash Rack/Tmnt)	Tacoma
7	OK	Tinker AFB (Soldier Cr/Bldg 3001)	Oklahoma City
7	CA	Lawrence Livermore Lab (USDOE)	Livermore
7	CA	Sharpe Army Depot	Lathrop
9	CA	Norton Air Force Base	San Bernardino
9	CA	Castle Air Force Base	Merced
10	NJ	Fort Dix (Landfill Site)	Pemberton
10	AL	Alabama Army Ammunition Plant	Childersburg
12	PA	Letterkenny Army Depot (SE Area)	Chambersburg
12	NY	Griffiss Air Force Base	Rome
12	VA	Defense General Supply Center	Chesterfield
12	WA	Fort Lewis (Landfill No. 5)	Tacoma
13	NM	Twin Cities AF (SAE Lndfill)	Minneapolis
13	MO	Lake City Army Plant (NW Lagoon)	Independence
13	IL	Joliet Army Ammu Plnt (Mfg Area)	Joliet
14	TX	Lone Star Army Ammu Plant	Texarkana
14	OR	Umatilla Army Depot (Lagoons)	Hermiston
15	WA	Bangor Ordnance Disposal	Bremerton
15	CA	Moffett Naval Air Station	Sunnyvale
16	CA	Mather AFB (AC&W Disposal Site)	Sacramento

* Sites are placed in groups (Gr) corresponding to groups of 50 on the final NPL.

Table A.12
CERCLA Extremely Hazardous Substances

Chemical Name	Notes	Threshold Planning Quantity (Pounds)
Acetone Cyanohydrin		1000
Acetone Thiosemicarbazide	e	1000/10000
Acrolein		500
Acrylamide	d,l	1000/10000
Acrylonitrile	d,l	10000
Acrylyl Chloride	e,h	100
Adiponitrile	e,l	1000
Aldicarb	c	100/10000
Aldrin	d	500/10000
Allyl Alcohol		1000
Allylamine	e	500
Aluminum Phosphide	b	500
Aminopterin	e	500/10000
Amiton	e	500
Amiton Oxalate	e	100/10000
Ammonia	l	500
Ammonium Chloroplatinate	a,e	10000
Amphetamine	e	1000
Aniline	d,l	1000
Aniline, 2,4,6-Trimethyl-	e	500
Antimony Pentafluoride	e	500
Antimycin A	c,e	1000/10000
ANTU		500/10000
Arsenic Pentoxide	d	100/10000
Arsenous Oxide	d,h	100/10000
Arsenous Trichloride	d	500
Arsine	e	100
Azinphos-Ethyl	e	100/10000
Azinphos-Methyl		10/10000
Bacitracin	a,e	10000
Benzal Chloride	d	500
Benzenamine, 3-(Trifluoromethyl)-	e	500
Benzene, 1-(Chloromethyl)-4-Nitro-	e	500/10000
Benzenearsonic Acid	e	10/10000
Benzenesulfonyl Chloride	a	10000
Benzimidazole, 4,5-Dichloro-2-(Trifluoromethyl)-	e,g	500/10000
Benzotrichloride	d	100
Benzyl Chloride	d	500
Benzyl Cyanide	e,h	500

Table A.12 (Continued)

Chemical Name	Notes	Threshold Planning Quantity (Pounds)
Bicyclo[2.2.1] Heptane-2-Carbonitrile, 5-Chloro-6-((((Methylamino)Carbonyl)Oxy)lmimo)-,(ls-(l-alpha,2-beta, 4-alpha, 5-alpha, 6E))-	e	500/10000
Bis(Chloromethyl) Ketone	e	10/10000
Bitoscanate	e	500/10000
Boron Trichloride	e	500
Boron Trifluoride	e	500
Boron Trifluoride compound with Methyl Ether (1:1)	e	1000
Bromadiolone	e	100/10000
Bromine	e,l	500
Butadiene	a,e	10000
Butyl Isovalerate	a,e	10000
Butyl Vinyl Ether	a,e	10000
Cadmium Oxide	e	100/10000
Cadmium Stearate	c,e	1000/10000
Calcium Arsenate	d	500/10000
Camphechlor	d	500/10000
Cantharidin	e	100/10000
Carbachol Chloride	e	500/10000
Carbamic Acid, Methyl-, 0-(((2,4-Dimethyl-l, 3-Dithiolan-2-yl) Methylene)Amino)-	e	100/10000
Carbofuran		10/10000
Carbon Disulfide	l	10000
Carbophenothion	e	500
Carvone	a,e	10000
Chlordane	d	1000
Chlorfenvinfos	e	500
Chlorine		100
Chlormephos	e	500
Chlormequat Chloride	e,h	100/10000
Chloroacetaldehyde	a	10000
Chloroacetic Acid	e	100/10000
Chloroethanol	e	500
Chloroethyl Chloroformate	e	1000
Chloroform	d,l	10000
Chloromethyl Ether	d,h	100
Chloromethyl Methyl Ether	c,d	100
Chlorophacinone	e	100/10000

Chemical Name	Notes	Threshold Planning Quantity (Pounds)
Chloroxuron	e	500/10000
Chlorthiophos	e,h	500
Chromic Chloride	e	1/10000
Cobalt	a,e	10000
Cobalt, ((2,2´-(1,2-Ethanediylbis (Nitrilomethylidyne))Bis(6-Fluorophenolato)) (2-)-N,N´,O,O´)-	e	100/10000
Cobalt Carbonyl	e,h	10/10000
Colchicine	e,h	10/10000
Coumafuryl	a,e	10000
Coumaphos		100/10000
Coumatetralyl	e	500/10000
Cresol, o-	d	1000/10000
Crimidine	e	100/10000
Crotonaldehyde		1000
Crotonaldehyde, (E)-		1000
Cyanogen Bromide		500/10000
Cyanogen Iodide	e	1000/10000
Cyanophos	e	1000
Cyanuric Fluoride	e	100
Cycloheximide	e	100/10000
Cyclohexylamine	e,l	10000
Cyclopentane	a,e	10000
C. l. Basic Green l	a,e	10000
Decaborane(14)	e	500/10000
Demeton	e	500
Demeton-S-Methyl	e	500
Dialifor	e	100/10000
Diborane	e	100
Dibutyl Phthalate	a	10000
Dichlorobenzalkonium Chloride	a,e	10000
Dichloroethyl Ether	d	10000
Dichloromethylphenylsilane	e	1000
Dichlorvos		1000
Dicrotophos	e	100
Diepoxybutane	d	500
Diethyl Chlorophospate	e,h	500
Diethylcarbamazine Citrate	e	100/10000
Diethyl-p-Phenylenediamine	a,e	10000
Digitoxin	c,e	100/10000
Diglycidyl Ether	e	1000
Digoxin	e,h	10/10000

Table A.12 (Continued)

Chemical Name	Notes	Threshold Planning Quantity (Pounds)
Dimefox	e	500
Dimethoate		500/10000
Dimethyl Phosphorochloridothioate	e	500
Dimethyl Phthalate	a	10000
Dimethyl Sulfate	d	500
Dimethyl Sulfide	e	100
Dimethyldichlorosilane	e,h	500
Dimethylhydrazine	d	1000
Dimethyl-p-Phenylenediamine	e	10/10000
Dimetilan	e	500/10000
Dinitrocresol		10/10000
Dinoseb		100/10000
Dinoterb	e	500/10000
Dioctyl Phthalate	a	10000
Dioxathion	e	500
Dioxolane	a,e	10000
Diphacinone	e	10/10000
Diphosphoramide, Octamethyl-		100
Disulfoton		500
Dithiazanine Iodide	e	500/10000
Dithiobiuret		100/10000
Emetine, Dihydrochloride	e,h	1/10000
Endosulfan		10/10000
Endothion	e	500/10000
Endrin		500/10000
Epichlorohydrin	d,l	1000
EPN	e	100/10000
Ergocalciferol	c,e	1000/10000
Ergotamine Tartrate	e	500/10000
Ethanesulfonyl Chloride, 2-Chloro-	e	500
Ethanol, 1.2-Dichloro-, Acetate	e	1000
Ethion		1000
Ethoprophos	e	1000
Ethylbis(2-Chloroethyl)Amine	e,h	500
Ethylene Fluorohydrin	c,e,h	10
Ethylene Oxide	d,l	1000
Ethylenediamine		10000
Ethyleneimine	d	500
Ethylmercuric Phosphate	a,e	10000
Ethylthiocyanate	e	10000
Fenamiphos	e	10/1000

Chemical Name	Notes	Threshold Planning Quantity (Pounds)
Fenitrothion	e	500
Fensulfothion	e,h	500
Fluenetil	e	100/10000
Fluorine	k	500
Fluoroacetamide	j	100/10000
Fluoroacetic Acid	e	10/10000
Fluoroacetyl Chloride	c,e	10
Fluorouracil	e	500/10000
Fonofos	e	500
Formaldehyde	d,l	500
Formaldehyde Cyanohydrin	e,h	1000
Formetanate Hydrochloride	e,h	500/10000
Formothion	e	100
Formparanate	e	100/10000
Fosthietan	e	500
Fuberidazole	e	100/10000
Furan		500
Gallium Trichloride	e	500/10000
Hexachlorocyclopentadiene	d,h	100
Hexachloronaphthalene	a,e	10000
Hexamethylenediamine, N,N´-Dibutyl-	e	500
Hydrazine	d	1000
Hydrocyanic Acid		100
Hydrogen Chloride (gas only)	e,l	500
Hydrogen Fluoride		100
Hydrogen Peroxide (Conc >52%)	e,l	1000
Hydrogen Selenide	e	10
Hydrogen Sulfide	l	500
Hydroquinone	l	500/10000
Indomethacin	a,e	10000
Iridium Tetrachloride	a,e	10000
Iron, Pentacarbonyl-	e	100
Isobenzan	e	100/10000
Isobutyronitrile	e,h	1000
Isocyanic Acid, 3,4-Dichlorophenyl Ester	e	500/10000
Isodrin		100/10000
Isofluorphate	c	100
Isophorone Diisocyanate	b,e	100
Isopropyl Chloroformate	e	1000

Table A.12 (Continued)

Chemical Name	Notes	Threshold Planning Quantity (Pounds)
Isopropyl Formate	e	500
Isoproplymethylpyrazolyl Dimethylcarbamate	e	500
Lactonitrile	e	1000
Leptophos	e	500/10000
Lewisite	c,e,h	10
Lindane	d	1000/10000
Lithium Hydride	b,e	100
Malononitrile		500/10000
Manganese, Tricarbonyl Methylcyclopentadienyl	e,h	100
Mechlorethamine	c,e	10
Mephosfolan	e	500
Mercuric Acetate	e	500/10000
Mercuric Chloride	e	500/10000
Mercuric Oxide	e	500/10000
Mesitylene	a,e	10000
Methacrolein Diacetate	e	1000
Methacrylic Anhydride	e	500
Methacrylonitrile	h	500
Methacryloyl Chloride	e	100
Methacryloyloxethyl Isocyanate	e,h	100
Methamidophos	e	100/10000
Methanesulfonyl Fluoride	e	1000
Methidathion	e	500/10000
Methiocarb		500/10000
Methomyl	h	500/10000
Methoxyethylmercuric Acetate	e	500/10000
Methyl 2-Chloroacrylate	e	500
Methyl Bromide	l	1000
Methyl Chloroformate	d,h	500
Methyl Disulfide	e	100
Methyl Hydrazine		500
Methyl Isocyanate	f	500
Methyl Isothiocyanate	b,e	500
Methyl Mercaptan		500
Methyl Phenkapton	e	500
Methyl Phosphonic Dichloride	b,e	100

Chemical Name	Notes	Threshold Planning Quantity (Pounds)
Methyl Thiocyanate	e	10000
Methyl Vinyl Ketone	e	10
Methylmercuric Dicyanamide	e	500/10000
Methyltrichlorosilane	e,h	500
Metolcarb	e	100/10000
Mevinphos		500
Mexacarbate		500/10000
Mitomycin C	d	500/10000
Monocrotophos	e	10/10000
Muscimol	a,h	10000
Mustard Gas	e,h	500
Nickel	a,d	10000
Nickel Carbonyl	d	1
Nicotine	c	100
Nicotine Sulfate	e	100/10000
Nitric Acid		1000
Nitric Oxide	c	100
Nitrobenzene	l	10000
Nitrocyclohexane	e	500
Nitrogen Dioxide		100
Nitrosodimethylamine	d,h	1000
Norbormide	e	100/10000
Organorhodium Complex (PMN-82-147)	e	10/10000
Orotic Acid	a,e	10000
Osmium Tetroxide	a	10000
Ouabain	c,e	100/10000
Oxamyl	e	100/10000
Oxetane, 3,3-Bis(Chloromethyl)-	l	500
Oxydisulfoton	e,h	500
Ozone	e	100
Paraquat	e	10/10000
Paraquat Methosulfate	e	10/10000
Parathion	c,d	100
Parathion-Methyl	c	100/10000
Paris Green	d	500/10000
Pentaborane	e	500
Pentachloroethane	a,d	10000
Pentachlorophenol	a,d	10000
Pentadecylamine	e	100/10000
Peracetic Acid	e	500
Perchloromethylmercaptan		500

Table A.12 (Continued)

Chemical Name	Notes	Threshold Planning Quantity (Pounds)
Phenol		500/10000
Phenol, 2,2´-Thiobis(4,6-Dichloro-	e	100/10000
Phenol, 2,2´-Thiobis(4-Chloro-6-Methyl-Phenol, 2,2´-Thiobis(4-Chloro-6-Methyl)-	e	100/10000
Phenol, 3-(1-Methylethyl)-, Methylcarbamate	e	500/10000
Phenoxarsine, 10, 10´-Oxydi-	e	500/10000
Phenyl Dichloroarsine	d,h	500
Phenylhydrazine Hydrochloride	e	1000/10000
Phenylmercury Acetate		500/10000
Phenylsilatrane	e,h	100/10000
Phenylthiourea		100/10000
Phorate		10
Phosacetim	e	100/10000
Phosfolan	e	100/10000
Phosgene	l	10
Phosmet	e	10/10000
Phosphamidon	e	100
Phosphine		500
Phosphonothioic Acid, Methyl-, O-Ethyl O-(4-(Methylthio)Phenyl) Ester	e	500
Phosphonothioic Acid, Methyl-, S-(2-(Bis(1-Methylethyl)Amino) Ethyl O-Ethyl Ester	e	100
Phosphonothioic Acid, Methyl-, O-(4-Nitrophenyl) O-Phenyl Ester	e	500
Phosphoric Acid, Dimethyl 4-(Methylthio) Phenyl Ester	e	500
Phosphorothioic Acid, O,O-Dimethyl-S-(2-Methylthio) Ethyl Ester	c,e,g	500
Phosphorus	b,h	100
Phosphorus Oxychloride	d	500
Phosphorus Pentachloride	b,e	500
Phosphorus Pentoxide	b,e	10
Phosphorus Trichloride		1000
Phylloquinone	a,e	10000
Physostigmine	e	100/10000
Physostigmine, Salicylate (1:1)	e	100/10000
Picrotoxin	e	500/10000

Chemical Name	Notes	Threshold Planning Quantity (Pounds)
Piperidine	e	1000
Piprotal	e	100/10000
Pinmifos-Ethyl	e	1000
Platinous Chloride	a,e	10000
Platinum Tetrachloride	a,e	10000
Potassium Arsenite	d	500/10000
Potassium Cyanide	b	100
Potassium Silver Cyanide	b	500
Promecarb	e,h	500/10000
Propargyl Bromide	e	10
Propiolactone, Beta-	e	500
Propionitrile		500
Propionitrile, 3-Chloro-		1000
Propiophenone, 4-Amino-	e,g	100/10000
Propyl Chloroformate	e	500
Propylene Glycol, Allyl Ether	a,e	10000
Propylene Oxide	l	10000
Propyleneimine	d	10000
Prothoate	e	100/10000
Pseudocumene	a,e	10000
Pyrene	c	1000/10000
Pyridine, 2-Methyl-5-Vinyl-	e	500
Pyridine, 4-Amino-	h	500/10000
Pyridine, 4-Nitro-, 1-Oxide	e	500/10000
Pyriminil	e,h	100/10000
Rhodium Trichloride	a,e	10000
Salcomine	e	500/10000
Sarin	e,h	10
Selenious Acid		1000/10000
Selenium Oxychloride	e	500
Semicarbazide Hydrochloride	e	1000/10000
Silane, (4-Aminobutyl)Diethoxymethyl-	e	1000
Sodium Anthraquinone-1-Sulfonate	a,e	10000
Sodium Arsenate	d	1000/10000
Sodium Arsenite	d	500/10000
Sodium Azide (Na(N3))	b	500
Sodium Cacodylate	e	100/10000
Sodium Cyanide (Na(CN))	b	100
Sodium Fluoroacetate		10/10000
Sodium Pentachlorophenate	e	100/10000

Table A.12 (Continued)

Chemical Name	Notes	Threshold Planning Quantity (Pounds)
Sodium Selenate	e	100/10000
Sodium Selenite	h	100/10000
Sodium Tellurite	e	500/10000
Stannane, Acetoxytriphenyl-	e,g	500/10000
Strychnine	c	100/10000
Strychnine, Sulfate	e	100/10000
Sulfotep		500
Sulfoxide, 3-Chloropropyl Octyl	e	500
Sulfur Dioxide	e,l	500
Sulfur Tetrafluoride	e	100
Sulfur Trioxide	b,e	100
Sulfur Acid		1000
Tabun	c,e,h	10
Tellurium	e	500/10000
Tellurium Hexafluoride	e,k	100
TEPP		100
Terbufos	e,h	100
Tetraethyllead	c,d	100
Tetraethyltin	c,e	100
Tetramethyllead	c,e,l	100
Tetranitromethane		500
Thallic Oxide	a	10000
Thallium Sulfate	h	100/10000
Thallous Carbonate	c,h	100/10000
Thallous Chloride	c,h	100/10000
Thallous Malonate	c,e, h	100/10000
Thallous Sulfate		100/10000
Thiocarbazide	e	1000/10000
Thiocyanic Acid, 2-(Benzothiazolythio)Methyl Ester	a,e	10000
Thiofanox		100/10000
Thiometon	a,e	10000
Thionazin		500
Thiophenol		500
Thiosemicarbazide		100/10000
Thiourea, (2-Chlorophenyl)-		100/10000
Thiourea, (2,Methylphenyl)-	e	500/10000
Titanium Tetrachloride	e	100
Toluene 2,4-Diisocyanate		500
Toluene 2,6-Diisocyanate		100
Trans-1,4-Dichlorobutene	e	500

Chemical Name	Notes	Threshold Planning Quantity (Pounds)
Triamiphos	e	500/10000
Triazofos	e	500
Trichloroacety Chloride	e	500
Trichloroethylsilane	e,h	500
Trichloronate	e,k	500
Trichlorophenylsilane	e,h	500
Trichlorophon	a	10000
Trichloro(Chloromethyl)Silane	e	100
Trichloro(Dichlorophenyl)Silane	e	500
Triethoxysilane	e	500
Trimethylchlorosilane	e	1000
Trimethylolpropane Phosphite	e,h	100/10000
Trimethyltin Chloride	e	500/10000
Triphenyltin Chloride	e	500/10000
Tris(2,Chloroethyl)Amine	e,h	100
Valinomycin	c,e	1000/10000
Vanadium Pentoxide		100/10000
Vinyl Acetate Monomer	d,l	1000
Vinylnorobornene	a,e	10000
Warfarin		500/10000
Warfarin Sodium	e,h	100/10000
Xylylene Dichloride	e	100/10000
Zinc, Dichloro(4,4-Dimethyl-5((((Methylamino)Carbonyl)Oxy) lmino)Pentanenitrile)-,(T-4)-	e	100/10000
Zinc Phosphide	b	500

a This chemical does not meet acute toxicity criteria. Its Threshold Planning Quantity (TPQ) is set at 10000 pounds.

b This material is a reactive solid. The TPQ does not default to 10000 pounds for non-powder, non-molten, non-solution form.

c The calculated TPQ changed after technical review as described in the technical support document.

d Indicates that the Reportable Quantity (RQ) is subject to change when the assessment of potential carcinogenicity and/or other toxicity is completed.

Table A.12 (Continued)

e Statutory reportable quantity for purposed of notification under Superfund Amendments & Reauthorization Act (SARA) sec 304(a)(2).

f The statutory 1 pound reportable quantity for methyl isocyanate may be adjusted in a future rulemaking action.

g New chemicals added that were not part of the original list of 402 substances.

h Revised TPQ based on new or re-evaluated toxicity data.

j TPQ is revised to its calculated value and does not change due to technical review as in proposed rule.

k The TPQ was revised after proposal due to calculation error.

l Chemicals on the original list that do not meet toxicity criteria but because of their high production volume and recognized toxicity are considered chemicals of concern ("Other Chemicals").

Table A.13
Hazardous Substances Prioritized for Further Study*

Priority Group 1 Substances	
-Benzo(a)pyrene	-Bis(2-ethylhexyl) phthalate
-Dibenzo(a),h)anthracene	-Tetrachloroethene
-Benzo(a)anthracene	-Benzo(b)fluoranthene
-Cyanide	-Chrysene
-Dieldrin/aldrin	-P-Dioxin
-Chloroform	-Lead
-Benzene	-Nickel
-Vinyl Chloride	-Arsenic
-Methylene chloride	-Beryllium
-Heptachlor/ heptachlor epoxide	-Cadmium
-Trichloroethane	-Chromium
-N-nitrosodiphenylamine	-PCB-1260,54,48,42,32,21, 1016
-1,4-Dichlorobenzene	
Priority Group 2 Substances	
-Carbon tetrachloride	-Benzidine
-Chlordane	-1,2-Dichloroethane
-N-nitrosodimethylamine	-Toluene
-4,4´-DDE,DDT,DDD	-Phenol
-Chloroethane	-Bis(2-chloroethyl)ether
-Bromodichloromethane	-2,4-Dinitrotoluene
-1,1-Dichloroethene	-BHC-alpha,gamma,beta, delta
-Isophorone	-Bis(chloromethyl)ether
-1,2-Dichloropropane	-N-nitrosodi-n-propylamine
-1,1,2-Trichloroethane	-Mercury
-1,1,2,2-Tetrachloroethane	-Zinc
-Pentachlorophenol	-Selenium
-3,3´-Dichlorobenzidine	

Table A.13 (Continued)

Priority Group 3 Substances

-1,1,1-Trichloroethane
-Chloromethane
-Oxirane
-Bromoform
-1,1-Dichloroethane
-Di-N-butyl phthalate
-2,4,6-Trichlorophenol
-Napthalene
-Nitrobenzene
-Ethylbenzene
-Acrolein
-Acrylonitrile
-Chlorobenzene
-Hexachlorobenzene
-1,2-Diphenylhydrazine
-Chlorodibromomethane
-1,2-Trans-dichloroethene
-Indeno(1,2,3-cd)pyrene
-2,6,Dinitrotoluene
-Total xylenes
-Endrin aldehyde/endrin
-Silver
-Copper
-Ammonia
-Toxaphene

Priority Group 4 Substances

-2,4-Diitrophenol
-P-Chloro-m-cresol
-Aniline
-Benzoic acid
-Hexachloroethane
-Bromomethane
-Carbondisulfide
-Fluorotrichloromethane
-Dichlorodifluoromethane
-2-Butanone
-Diethyl phthalate
-Phenanthrene
-Hexachlorobutadiene
-Phenol,2-methyl
-1,2-Dichlorobenzene
-2,4-Dimethylphenol
-2-Pentanone, 4-methyl
-1,2,4-Trichlorobenzene
-2,4-Dichlorophenol
-1,4-Dioxane
-Dimethyl phthalate
-Fluoranthene
-4,6-Dinitro-2-methylphenol
-1,3-Dichlorobenzene
-Thallium

* The four groups are listed in descending order of priority, with the first group having the highest priority substances of the first priority list. The substances within each group are listed in Chemical Abstracts Services (CAS) number order. The first (and highest) priority group of 25 hazardous substances is composed of the substances which will be the subject of the first toxicological profiles developed under section 110 of Superfund Amendments & Reauthorization Act (SARA).

Table A.14
Asbestos-Containing Materials Found in Buildings

Subdivision	Generic Name	Asbestos (%)	Dates of use	Binder/Sizing
Surfacing material	sprayed- or troweled-on	1-95	1935-1970	Sodium silicate, portland cement, organic binders
Preformed thermal insulating products	batts, blocks, & pipe covering			
	85% magnesia	15	1926-1949	magnesium carbonate
	calcium silicate	6-8	1949-1971	calcium silicate
Textiles	cloth**			
	blankets (fire)**	100	1910-present	none
	felts:	90-95	1920-present	cotton/wool
	blue stripe	80	1920-present	cotton
	red stripe	90	1920-present	cotton
	green stripe	95	1920-present	cotton
	sheets	50-95	1920-present	cotton/wool
	cord/rope/yarn**	80-100	1920-present	cotton/wool
	tubing	80-85	1920-present	cotton/wool
	tape/strip	90	1920-present	cotton/wool
	curtains** (theater, welding)	60-65	1945-present	cotton
Cementitious concrete-like products	extrusion panels:	8	1965-1977	portland cement
	corrugated	20-45	1930-present	portland cement
	flat	40-50	1930-present	portland cement
	flexible	30-50	1930-present	portland cement

Table A.14 (Continued)

Subdivision	Generic Name	Asbestos (%)	Dates of use	Binder/Sizing
Cementitious concrete-like products (con.)	flexible perforated	30-50	1930-present	portland cement
	laminated (outer surface)	35-50	1930-present	portland cement
	roof tiles	20-30	1930-present	portland cement
	clapboard & shingles:			
	clapboard	12-15	1944-1945	portland cement
	siding shingles	12-14	?-present	portland cement
	roofing shingles	20-32	?-present	portland cement
	pipe	20-15	1935-present	portland cement
Paper products	corrugated:			
	high temperature	90	1935-present	sodium silicate
	moderate temperature	35-70	1910-present	starch
	indented	98	1935-present	cotton & organic binder
	millboard	80-85	1925-present	starch, lime, clay
Roofing felts	smooth surface	10-15	1910-present	asphalt
	mineral surface	10-15	1910-present	asphalt
	shingles	1	1971-1974	asphalt
	pipeline	10	1920-present	asphalt
Asbestos-containing compounds	caulking putties	30	1930-present	linseed oil
	adhesive (cold applied)	5-25	1945-present	asphalt
	joint compound		1945-1975	asphalt
	roofing asphalt	5	?-present	asphalt
	mastics	5-25	1920-present	asphalt

Subdivision	Generic Name	Asbestos (%)	Dates of use	Binder/Sizing
Asbestos-containing compounds (con.)	asphalt tile cement	13-25	1959-present	asphalt
	roof putty	10-25	?-present	asphalt
	plaster/stucco	2-10	?-present	portland cement
	spackles	3-5	1930-1975	starch, casein, synthetic resins
	sealants fire/water	50-55	1935-present	caster oil or poly-isobutylene
	cement, insulation	20-100	1900-1973	clay
	cement, finishing	55	1920-1973	clay
	cement, magnesia	15	1926-1950	magnesium carbonate
Asbestos ebony products		50	1930-present	portland cement
Flooring tile and sheet goods	vinyl/asbestos tile	21	1950-present	poly(vinyl)chloride
	asphalt/asbestos tile	26-33	1920-present	asphalt
	sheet goods/resilient	30	1950-present	dry oils
Wallcovering	vinyl wallpaper	6-8	?-present	--
Paints and coatings	roof coating	4-7	1900-present	asphalt
	air tight	15	1940-present	asphalt

* The information in this Table is taken, with modification, from: Lory EE, Coin DC.February 1981. Management Procedure for Assessment of Friable Asbestos Insulating Material. Port Hueneme, CA: Civil Engineering Laboratory Naval Construction Battalion Center. The U.S. Navy prohibits the use of asbestos-containing materials when acceptable nonasbestos substitutes have been identified.

** Laboratory aprons, gloves, cord, rope, fire blankets, and curtains may be common in schools.

? Unknown beginning date of use.

Table A.15
Asbestos Abatement Methods for Surfacing Materials

Method/ Applicability	Advantages	Disadvantages	General Comments
Removal/ can be used in most situations	Eliminates asbestos source without requiring special operations or maintenance program	Replacement with substitute material may be necessary as well as encapsulation of porous surfaces Improper removal may raise fiber levels	Containment barriers and worker protection required Wet removal is required for all types of asbestos Disposal may be a problem in some areas Unusual circumstances, complex surfaces, and presence of utilities may require special removal techniques
Enclosure/ ACM located in a small area (e.g., a column) Disturbance or entry into enclosed area unlikely	Reduces exposure outside of enclosure Initial costs may be lower than for removal unless utilities need relocating or major changes	Asbestos source continues releasing fibers within the enclosed area and must must be removed eventually Special operations program to control access to enclosure for maintenance and renovation required	Containment barriers needed Inappropriate for damaged or deteriorating materials, where entry into enclosure likely,

Method/ Applicability	Advantages	Disadvantages	General Comments
Enclosure (cont)	Usually does not require replacement of material	Periodic reinspection required to check for damage	or ceiling to be enclosed is low
		Repair of damaged enclosure necessary	
		Fibers released in dry form during construction of the enclosure	
		Long term costs could be higher than for removal	
Encapsulation/material retaining bonding integrity, not highly accessible	Reduces asbestos fiber release from material	Asbestos source remains and must be removed later	Containment barriers and worker protection needed
	Initial costs may be lower than removal	If material is not in good in condition, sealant may cause material to delaminate	Airless sprayers should be used
Material granular, cementitious	Does not require replacement of material	Periodic reinspection to check for damage or deterioration or required	Previously encapsulated materials may

Table A.15 (Continued)

Method/ Applicability	Advantages	Disadvantages	General Comments
Encapsulation/ after removal of ACM, if the substrate is porous		Repair of damaged or deteriorating encapsulated surfaces required	have to be re-encapsulated
		Encapsulated surface is difficult to remove and may require dry techniques for eventual removal	Inappropriate for material which is damaged, does not adhere to substrate, or is fibrous, fluffy
		Long-term costs may be higher than removal	

Table A.16
Some Incompatible Chemicals

Chemical	Incompatible Chemicals
Acetic acid	Chromic acid, ethylene glycol, hydroxyl-containing compounds, nitric acid, perchloric acid, permanganates, and peroxides
Acetone	Bromine, chlorine, nitric acid, and sulfuric acid
Acetylene	Bromine, chlorine, copper, silver, fluorine, and mercury
Alkaline and alkaline earth metals such as powdered calcium, cesium, lithium, magnesium, potassium, sodium, aluminum, etc.	Carbon dioxide, chlorinated hydrocarbons, water, and the halogens
Aluminum and its alloys (particularly powders)	Acid or alkaline solutions, water, ammonium persulfate, chlorates, chlorinated compounds, nitrates, and organic compounds in nitrate/nitrite salt baths
Ammonia (anhydrous)	Bromine, calcium hypochlorite, chlorine, hydrofluoric acid, iodine, mercury, and silver
Ammonium nitrate	Acids, chlorates, chlorides, lead, metallic nitrates, metal powders, finely divided organics or combustibles, sulfur, and zinc
Ammonium perchlorate, permanganate, or persulfate	Combustible materials, oxidizing materials such as acids, chlorates and nitrates
Aniline	Hydrogen peroxide or nitric acid
Barium peroxide	Combustible organics, oxidizing materials, water, and reducing agents

Table A.16 (Continued)

Chemical	Incompatible Chemicals
Bismuth and its alloys	Perchloric acid
Bromine	Acetone, acetylene, ammonia, benzene, butadiene, butane and other petroleum gases, hydrogen, finely divided metals, sodium carbide, and turpentine
Calcium or sodium carbide	Moisture (in air) or water
Calcium hypochlorite	(Activated) ammonia or carbon containing materials
Carbon, activated	Calcium hypochlorite and all oxidizing agents
Chlorates or perchlorates	Acids, aluminum, ammonium salts, cyanides, phosphorous, metal powders, finely divided organics or other combustibles, sugar, sulfides, and sulfur
Chlorine	Same as bromine
Chlorine dioxide	Ammonia, hydrogen, sulfide, methane and phosphine, organic materials
Chromic acid	Acetic acid (glacial), acetic anhydride, alcohols, combustible materials, flammable liquids, glycerin, naphthalene, nitric acid, sulfur, turpentine, reducing agents, and oxidizing materials
Cooper	Acetylene, hydrogen peroxide, sodium azide
Cumene hydroperoxide	Acids (mineral or organic)
Cyanides	Acids or alkalies

Chemical	Incompatible Chemicals
Flammable liquids	Ammonium nitrate, chromic acid, hydrogen peroxide, nitric acid, sodium peroxide, and the halogens
Fluorine	Most materials
Hydrocarbons such as benzene, butane, gasoline, propane, turpentine, etc.	Bromine, chlorine, chromic acid, fluorine, hydrogen peroxide, and sodium peroxide
Hydrofluoric acid or anhydrous hydrogen fluoride	Ammonia (anhydrous or aqueous)
Hydrocyanic acid or cyanide	Alkalies, nitric acid, oxidizers
Hydrogen peroxide 3%	Chromium, copper, iron, most metals or their salts
Hydrogen peroxide 30% to 90%	Same as 3% hydrogen peroxide plus aniline, any flammable liquids, combustible materials, nitromethane, and all other organic matter, including alcohols
Hydrogen sulfide	Fuming nitric acid or oxidizing gases
Iodine	Acetylene, ammonia (anhydrous or aqueous), and hydrogen-reducing materials
Lithium	Acids, moisture in air, water, oxidizers
Lithium aluminum hydride	Acids, chlorinated hydrocarbons, carbon dioxide, ethyl acetate, and water-powdered limestone as extinguishing agent
Magnesium (particularly powder)	Carbonates, chlorates, heavy metal oxalates or oxides, nitrates, perchlorates, peroxides, phosphates, and sulfates

Table A.16 (Continued)

Chemical	Incompatible Chemicals
Mercuric oxide	Sulfur, reducing agents
Mercury	Acetylene, alkali metals, ammonia, nitric acid with ethanol, oxalic acid, azides
Nitrates	Combustible and flammable materials, esters, phosphorous, sodium acetate, stannous chloride, water and zinc powder
Nitric acid (concentrate)	Acetic acid, aniline, chromic acid, flammable gases and liquids, hydrocyanic acid and hydrogen sulfide
Nitric acid	Alcohols and other oxidizable organic material, hydriodic acid (hydrogen iodide) magnesium or other metals, phosphorous, and thiophene
Nitrites	Potassium or sodium cyanide, reducing agents
Oxalic acid	Mercury or silver
Oxygen (liquid or enriched air)	Flammable gases, liquids, or solids such as acetone, acetylene, grease, hydrogen, oils, and phosphorous
Perchloric acid	Acetic anhydrides, alcohols, bismuth and its alloys, grease, oils or any organic materials, and reducing agents, paper, wood
Peroxides (organic)	Acids (mineral or organic), reducing agents
Phosphorous	Chlorates and perchlorates, nitric acid and nitrates, organic materials, reducing agents, white phosphorous, air or oxygen in addition to others

Chemical	Incompatible Chemicals
Phosphorous pentoxide	Organic compounds, water, reducing agents
Picric acid	Ammonia heated with oxides or salts or heavy metals (particularly copper, lead, zinc) and friction with oxidizing agents and reducing agents
Potassium	Air (moisture and/or oxygen), water, carbon dioxide, carbon tetrachloride
Potassium chlorate or perchlorate	Acids or their vapors, combustible materials, especially organic solvents, phosphorous, and sulfur
Potassium permanganate	Benzaldehyde, ethylene glycol, glycerin, and sulfuric acid
Silver	Acetylene, ammonium compounds, nitric acid with ethanol, oxalic acid, tartaric acid, fulminic acid, azides
Sodium	Carbon tetrachloride, carbon dioxide, and water
Sodium amide	Air (moisture and oxygen), water, oxidizers
Sodium chlorate	Acids, ammonium salts, oxidizable materials and sulfur
Sodium hydrosulfite	Air (moisture) or combustible materials
Sodium nitrite	Ammonia compounds, ammonium nitrate or other ammonium salts, organic materials, friction
Sodium peroxide	Acetic acid (glacial), acetic anhydride, alcohols, benzaldehyde, carbon disulfide, ethyl acetate, ethylene glycol, furfural,

Table A.16 (Continued)

Chemical	Incompatible Chemicals
Sodium peroxide (continued)	glycerin, methyl acetate, and other oxidizable substances, powdered metals, water, acids, organic materials
Sulfur	Any oxidizing materials
Sulfuric acid	Chlorates, perchlorates, and permanganates, combustibles
Water	Acetyl chloride, carbides, chromic acid, phosphorous oxychloride, phosphorous pentechloride, sulfuric acid, and sulfur trioxide
Zinc chlorate	Acids or organic materials
Zinc (particularly powder)	Acids or water
Zirconium (particularly in powder form)	Carbon tetachloride and other halogenated hydrocarbons, peroxides, sodium bicarbonate, water, and air

Source: Safety Engineering Standards, Industrial Indemnity is the sole owner of the rights granted and Industrial Indemnity controls these rights.

GLOSSARY

With each new development and as each new discipline has come into being, there has been an expansion of the vocabulary used by workers in these areas . . .

—Daniel N. Lapedes, Editor-in-Chief
McGraw-Hill Book Company, 1974

Abatement. The method of reducing the degree or intensity of pollution, also the use of such a method.

Absorption. The taking up of matter in bulk by other matter.

Accuracy. The extent to which the results of a calculation or the readings of an instrument approach the true values of the calculated or measured quantities, and are free from error.

Acid Waste. A waste with a pH less than 7. An acid waste is hazardous when its pH is 2 or less.

Action Level. The exposure concentration at which certain provisions of the NIOSH recommended standards must be initiated, such as periodic measurements of worker exposure, training of workers, and medical surveillance.

Activated Carbon. A highly absorbent form of carbon, used to remove odors and toxic substances from gaseous emissions. In advanced waste treatment, activated carbon is used to remove dissolved organic matter from waste water.

Acute. Referring to a disease or disorder of rapid onset, short duration, and pronounced symptoms. In exposure, concentrated and of short duration.

Administrative Procedure Act. The controlling statute for rule making.

Adsorption. The surface retention of solid, liquid, or gas molecules, atoms, or ions by a solid or liquid, as opposed to absorption, the penetration of substances into the bulk of the solid or liquid.

Aerated Lagoon. A body of water where air is introduced by spraying, stirring, or a similar method of agitation, in order to promote microbiological degradation of waste substances.

Aerobic. Processes that can occur only in the presence of oxygen.

Agreement State. A state which has assumed regulatory authority from the Nuclear Regulatory Commission over radioactive by-products, source materials, and small quantities of special nuclear materials.

Air Quality Control Region. An area designated by the federal government where two or more communities, either in the same or different states, share a common air pollution problem or common air quality conditions.

Alkaline Waste. A waste with a pH between 7 and 14. A waste is hazardous when its pH is 12.5 or greater.

Anaerobic. Processes that occur in the absence of oxygen.

Anaerobic Digestor. A vessel in which microorganisms decompose wastes in the absence of air.

Anesthetic. Material capable of causing a marked loss of muscular powers.

Aquifer. A subsurface zone that yields economically important amounts of water to wells or springs, usually composed of gravel or porous stone.

Asbestos. A generic name given to a number of naturally occurring hydrated mineral silicates that possess a unique crystalline structure, are incombustible in air, and are separable into fibers.

Asbestos Abatement. Procedures to control fiber release from asbestos-containing materials in buildings.

Asbestosis. A nonmalignant, progressive, irreversible lung disease caused by the inhalation of asbestos dust and characterized by diffuse fibrosis.

Asphyxiant. Material that is capable of arresting respiration.

Auto-Ignition Temperature. Temperature at which ignition occurs without the introduction of an ignition source. (Same as spontaneous combustion.)

Biological Treatment. Process in which living microorganisms are mixed with or come in contact with waste, and use the biologically degradable organics in the waste as an energy source, converting the waste to methane, oxygen, or water. Distinguished from physical/chemical methods, which do not employ living microorganisms.

By-product. A material produced without separate commercial intent during the manufacture, processing, or treatment of other materials or mixtures.

Carcinogen. Any agent that incites development of a carcinoma or any other sort of malignancy.

cc. Cubic centimeter, a measure of volume.

Ceiling. A description usually seen in connection with a published exposure limit, referring to a concentration that should not be exceeded, even for an instant.

Centrifugation. The use of a mechanical device which separates solids from liquids or separates components of liquid emulsions through use of centrifugal force.

Characteristics of Hazardous Wastes. A method of identifying which substances are hazardous waste by their physical/chemical properties. EPA has defined four "characteristics" that can be determined by tests:

1. Ignitability: the ability to catch fire.
2. Corrosivity: the ability to corrode other materials.
3. Reactivity: the ability to enter into a violent chemical reaction, which may involve explosion or fumes.
4. EP (Extraction Procedure) Toxicity: the ability to release certain toxic constituents when leached with mild acid.

Chemical Dechlorination. The stripping of chlorine atoms from highly chlorinated toxic compounds.

Chronic. Long-continued; of long duration.

Ci. Curie, a unit of radioactivity.

cm. Centimeter, a measure of length.

Coliform Organism. Any of a number of organisms common to the intestinal tract of man and animals whose presence in waste water is an indicator of pollution and of potentially dangerous bacterial contamination.

Compliance. The act of conforming with a demand or proposal, as in complying with regulatory or permit requirements.

Confidential Business Information. Any information in any form received by EPA from any person, firm, partnership, corporation, association or local, state, or federal agency which contains trade secrets, formulas, commercial or financial information, and which has been claimed as confidential by the person submitting it and which has not been legally determined to be nonconfidential by the EPA General Counsel.

Contingency Planning. Preparing an emergency procedure before an actual emergency happens.

Corrosive Material. One that undergoes a chemical process which converts minerals and metals into unwanted products.

Criteria Pollutants. Selected and specified polluting materials for which regulatory limitations are established.

Cryogens. Gases that must be cooled to less than $-1500\,°F$ to achieve liquefaction.

Danger. The potential consequences of exposure to a hazard, including some measure of severity. The potential for injury to some living organism is usually implied.

DDT. The first of the modern chlorinated hydrocarbon insecticides whose chemical name is 1,1,1-tricholoro-2,2-bs (p-chloriphenyl)-ethane. Because of its persistence in the environment and its ability to accumulate and magnify in the food chain, EPA has banned the registration and interstate sale of DDT for nearly all uses in the United States, effective December 31, 1972.

Degradation. Chemical or biological transformation of a complex compound into a number of simple ones.

Determinate Error. Errors that can be ascribed to definite causes. Also known as systematic errors.

Dialysis. A membrane separation technology that uses osmosis as the primary driving force for transport of components from one side of the membrane to the other.

Discharge. The release of any waste stream or any constituent thereof into the environment.

Disposal. Final placement of waste.

Distillation. Using boiling to separate a liquid into components or to separate dissolved or trapped solids from a liquid into stream. Frequently used for recycling used solvents.

Domestic Waste. Solid waste, composed of garbage and rubbish, which normally originates in a residential private household or apartment house. Also known as household waste.

Dupe. Short for duplicate. One of two or more samples that contain the same materials in identical amounts.

Effluent. A discharge of pollutants into the environment, partially or completely treated or in its natural state. Generally used in regard to discharges into water.

Electrodialysis. Dialysis that is conducted with the aid of an electromotive force applied to electrodes adjacent to both sides of a membrane.

Enforcement. The act of compelling obedience or compliance.

Environmental Audit. A formal self-appraisal conducted to assess facility compliance with environmental problems and liabilities.

Environmental Impact Statement. A document prepared by EPA or under EPA guidance, which identifies and analyzes in detail the environmental impacts of a proposed action.

Error. The numerical difference between a measured value and the true value.

Evaporation. The process by which a liquid passes from a liquid state to a vaporous state below the boiling point of the liquid.

Explosive. A chemical compound or mixture of compounds which suddenly undergoes a very rapid chemical transformation, with the simultaneous production of large quantities of heat and gases.

Explosive Wastes. Wastes that are unstable and may readily undergo violent chemical change or explode.

Exposure. In risk assessment, a quantification of the populations at risk and the magnitude and duration of their exposure. Medically, the state of being open to some action or influence that may affect detrimentally, such as toxic chemicals or radioactivity.

Filtration. Using a porous material to separate solids from a liquid.

Fire Point. Temperature at which a material gives off enough vapor to continue to burn when lighted.

Fixation. See *solidification*.

Flammable Range. The numerical difference between the upper and lower explosive limits.

Flashpoint. Minimum temperature at which a spark or flame will cause an instantaneous flash in the vapor space above a material.

Flotation. Separation of solids from liquids using the attachment of tiny air bubbles to the solid particles.

Flue Gas Desulfurization. The operation of removing sulfur oxides from exhaust gas streams of a boiler or industrial process, results in a by-product commonly termed FGD sludge.

Fluid Bed Incinerator. Involves the use of a fluidizing material such as alumina which is fluidized by an air stream. The fluidized bed acts as a hearth.

Friable. Capable of being crumbled, pulverized, or reduced to powder by hand pressure.

Fungicide. A pesticide chemical that kills fungi or prevents them from causing disease, usually on plants of economic importance. See *pesticide*.

Garbage. Solid waste resulting from animal, grain, fruit, or vegetable matter used or intended for use as food. Also unwanted or used material.

Grate Type Incinerator. Refractory lined box utilizing either a stationary or moving grate onto which the waste is fed.

Groundwater. A body of water, generally within the boundaries of a watershed, which exists in the internal passageways of porous geological formations (aquifers) and which flows in response to gravitational forces.

Hazard. A condition that can be expected to cause damages, including injury or death, to exposed individuals.

Hazardous Materials. Any materials, including substances and wastes, that may pose an unreasonable risk to health, safety, property, or the environment, when they exist in specific quantities and forms.

Hazardous Substances. Hazardous substances are a subset of hazardous materials and are identified and regulated by EPA.

Hazardous Waste. Waste which, if improperly managed, can create a risk to the safety or health of people or to the environment. EPA considers hazardous waste a subset of both solid waste and hazardous materials.

Hearth Type Two-Chamber Incinerator. Utilizes principles of burning waste on a solid hearth with controlled forced air addition. Second chamber has additional air and an auxiliary burner.

Hydrolysis. The decomposition or alteration of a chemical substance by water.

Hygroscopic. Pertaining to a substance, whose physical characteristics are significantly altered by the effects of water vapor.

Immediately Dangerous to Life or Health. Exposure level for the purpose of respirator protection, representing a maximum concentration from which, in the event of respirator failure, one could escape within 30 minutes without experiencing any escape-impairing or irreversible health effects.

Immobilization. See *solidification.*

Incineration. The controlled burning of a waste at very high temperatures for a specified length of time to destroy combustible constituents in the waste.

Indeterminate Error. Errors that exist because of unpredictable and imperceptible factors, causing what appear to be random fluctuations in measurements.

Ion Exchange. A reversible exchange of ions (electrically charged atoms or groups of atoms) between a solid and a liquid in water and waste water treatment, the solids, ion-exchange resins, exchange the polluting ions in the water for ions originally on the resin.

Irritant. Material that produces local irritation or inflammation of eyes, skin, or respiratory membranes.

Irritating Material. A liquid or solid substance which, upon contact with fire or when exposed to air, gives off dangerous or intensely irritating fumes, but not including a poisonous material.

kg. Kilogram, a measure of mass.

Kindling Point. Minimum solid temperature at which sufficient vapors exist to burn with a flame.

Knowing Endangerment. Conduct that knowingly places another person in danger of death or bodily injury, or manifests an unjustifiable and inexcusable disregard for human life.

Label. A sticker of specified design that goes on a package or container and pictorially indicates the hazard class of the contents.

Land Farming. A waste treatment technique that uses microorganisms naturally occurring in soil to degrade waste materials.

Landfills. Excavations in which wastes are buried.

Law. A rule of conduct or action prescribed or formally recognized as binding or enforced by a controlling authority.

Lethal Concentration. The amount of a substance which results in death for 50 percent of test specimens after a specific duration of exposure, usually expressed in ppm by volume.

Lethal Dose. The amount of a substance which results in death for 50 percent of test specimens after a specific duration of exposure, measured generally as quantity per unit of body weight in mg/kg.

Listed Waste. Wastes listed by EPA as hazardous by definition, even in instances where the wastes may not exhibit the defined characteristics of hazardous waste.

Lower Explosive Limit. The minimum concentration of gas or vapor in air below which a substance does not burn when exposed to an ignition source.

Lung Cancer. An uncontrolled growth of abnormal cells in the lungs, which normally results in death.

m. Meter, a measure of length.

m^3. Cubic meter, a measure of volume.

Manifest System. A procedure required by EPA in which hazardous waste quantities are identified as they are produced and are tracked through treatment, transportation, and disposal by a series of permanent, linkable, descriptive documents.

Marking. Written information which identifies package or container contents.

Mechanical Integrity. A general standard for injection wells which signifies that there is no significant leakage in the well's casing, tubing, or packer; and no significant movement of fluids between the outermost casing and the well bore.

Mesothelioma. A relatively rare form of cancer which develops in the lining of the pleura or peritoneum with no known cure.

μg. Microgram, a unit of mass.

mg. Milligram, a unit of mass.

Monitoring. The measurement, sometimes continuous, of specific properties at a particular point to detect changes which may be attributed to a nearby operation and controlled by the operation.

Monitoring Well. A well, used to obtain water samples for water quality analysis or to measure groundwater levels.

Multiple Hearth Incinerator. Usually consists of four or more horizontal refractory-lined hearths arranged one above the other in a vertical configuration.

Mutagen. An agent that raises the frequency of mutation above the spontaneous rate.

Neutralization. The conversion of acid or alkaline substances to a state that is no longer highly corrosive.

Ocean Disposal. The depositing of waste into ocean or estuarine waters.

Ocean Dumping. The use of various techniques for disposing of wastes in open seas. Includes bulk disposal of liquid or slurry-type wastes, sludges, the sinking of containerized wastes, and burial of wastes in the sea bed.

Oxidation. A chemical reaction in which oxygen unites or combines with other elements.

Oxidizing Agent. A substance that gains electrons in a chemical reaction.

Permissible Exposure Limit. An exposure limit that is published and enforced by the Occupational Safety and Health Administration as a legal standard.

Permit. A document demonstrating the official approval and permission to proceed with an activity controlled by the permitting authority, usually a governmental agency. Several permits from different authorities may be required for a single operation.

Pesticide. Any substance or mixture of substances intended for preventing, destroying, repelling, or mitigating any pest, or for use as plant defoliant or dessicant. See *fungicide and rodenticide.*

pH. The measure of the acidity or alkalinity of a chemical solution given in a range from 0 to 14, derived from the reciprocal of the logarithm of the hydrogen ion concentration.

Pharmacology. The study of the nature, properties, and the effects that chemicals have on living systems.

Physical Treatment. Consists of nonchemical methods designed to reduce the volume of waste or to separate out a portion of the waste. It does not result in the destruction of the toxic components of the waste.

Placard. Similar to a label in that it pictorially illustrates the hazard class of material being shipped. Is placed on the transport vehicle rather than on the material container.

Point Source. Any discernible, confined, and discrete conveyance, including but not limited to any pipe, ditch, channel, tunnel, conduit, well, discrete fissure, container, rolling stock, concentrated animal feeding operation, or vessel or other floating craft, from which pollutants are or may be discharged. Does not include return flows from irrigated agriculture.

Polychlorinated Biphenyls. A group of organic compounds used in the manufacture of plastics, known by the acronym PCBs. In the environment, PCBs exhibit many of the same characteristics as DDT and may be confused with that pesticide. PCBs are highly toxic to aquatic life, they persist in the environment for a long time, and they are biologically accumulative.

Polymers. Generally organic compounds. Commonly composed of resins, plastics, and fibers.

ppb. Parts per billion, a unit of concentration.

ppm. Parts per million, a unit of concentration.

Precipitation. The chemical operation of causing a material dissolved in a fluid to react with another material to form solids that can be removed and can be managed separately.

Precision. The quality of being exactly or sharply defined or stated. In analyses, often the quality of being reproducible.

Priority Pollutants. A specific list of 129 chemicals selected from the list of toxic pollutants by the EPA as priority toxic pollutants for regulation under the Clean Water Act. See *toxic pollutants.*

Promulgate. To make known or put into force by open declaration.

Prospective. Dealing with future and existing activities, not with past events.

Pyrolysis. The chemical decomposition of a material by heat in the absence of oxygen.

Pyrophoric. Substances that spontaneously ignite in air without an ignition source.

Quality Assurance. Testing and inspecting of all or a portion of the final product to ensure that the desired quality level of product reaches the consumer.

Quality Control. Inspection, analysis, and action applied to a portion of the product or operation to estimate overall quality of the product and determine what, if any, changes must be made to achieve or maintain the required level of quality.

Radioactive Material. Any material, or combination of materials, that spontaneously emits ionizing radiation, and having a specific activity greater than 0.002 microcurie per gram.

Recommended Exposure Limit. An exposure limit recommended by the National Institute for Occupational Safety and Health, not to be exceeded.

Recycling. Commonly, the use of discarded materials and objects in original or changed form rather than wasting them. Precisely, the practice of sending a material back into the process by which it was first formed.

Reduction. The addition of hydrogen or the abstraction of oxygen from a substance.

Regulation. A rule or order having the force of a law, issued by an executive authority of government, such as an agency.

Rem. Roentgen equivalent man, a measure of the biological damage to living tissue from radiation exposure.

Replicate. See *dupe.*

Reverse Osmosis. A treatment technique in which pressure is applied to the surface of a waste solution, forcing pure water to pass from the solution through a selective membrane.

Risk. The probability of exposure, coupled with the severity of the consequences. Risk is often used in a more general way than *danger*, in that risk is used to describe potential financial loss or property damage in addition to environmental damage or personal injury.

Risk Assessment. A process that takes into account the available information about exposure to hazardous materials and assesses what data are still necessary to collect.

Rodenticide. A chemical or agent used to destroy or prevent damage by rats or other rodent pests. See *pesticide.*

Roentgen. A measure of the total amount of ionization that the quantity of radiation could produce in the air.

Rotary Kiln Incinerator. Large cylindrical kiln rotated on trunnions.

Runoff. Water that, having fallen on a ground surface (natural, paved, or otherwise), flows across the surface, picking up materials, and will, if not collected, continue into a watercourse or wetland.

Safety. Methods and techniques of avoiding accidents and disease, a judgment of the acceptability of risks.

Sample. Groups or portions of material, taken from a larger collection of units or quantity of material, that provide information to be used for judging the quality of the total collection or entire material as a basis for action.

Screening. The removal of relatively coarse floating and suspended solids by straining through racks or screens.

Sedimentation. Using gravity to separate solids from a liquid. Frequently used where the concentration of solids is quite low.

Short Term Exposure Limit. Concentration to which workers can be exposed for fifteen minutes for only four times throughout the day with at least one hour between exposures.

Silviculture. The development and care of forests.

Site. The property on which a facility is located.

Smoke. A gas which is the result of combustion and commonly acts both as an asphyxiant and an irritant.

Sole Source Aquifer. Underground water supplies that serve as the only source of drinking water for a specific area.

Solidification. The conversion of a liquid or semi-liquid substance into a solid mass. Intended to immobilize hazardous constituents, reducing the potential for them to migrate further into the environment. Also known as fixation, stabilization, immobilization.

Solid Waste. A term defined by EPA that may include nonsolid discarded materials.

Solvent Extraction. A process that removes organic substances by contacting them with a solvent that is immiscible with at least one component of the waste stream.

Somatic Cells. Cells containing similar but not identical pairs of chromosomes; thus, the great majority of body cells, but excluding germ cells.

Spike. A sample to which a known amount of a specific substance has been added.

Stabilization. See *solidification*.

Statute. A law enacted by the legislative branch of a government.

Storage. Temporary holding of a material pending utilization, transportation, treatment, or disposal, in such a manner as not to constitute disposal.

Strict Liability. Liability without fault.

Stripping. A process in which the most volatile components are removed from a mixture. Stream or air contact may be used.

Superfund. Hazardous Substance Response Trust Fund.

Synergism. When chemicals work together in such a way that they are far more toxic when mixed together than their simple additive effect could account for.

Threshold Dose. The minimum dose of a given substance necessary to produce a measurable physiological or psychological effect.

Threshold Limit Value. The upper limit of a toxicant concentration to which an average, healthy person may be exposed all day, every day, without suffering adverse effects.

Time Weighted Average. The average time, over a given work period, of a person's exposure to a chemical or an agent, determined by sampling for the contaminant throughout the time period.

Toxic. Capable of producing injury, illness, or damage to humans, domestic livestock, wildlife, or other organisms through ingestion, inhalation, or absorption through any body surface.

Toxicity. Relates to a harmful effect on the body by physical contact, ingesting, or inhalation. The property of being poisonous, capable of causing death or severe disability of an organism.

Toxic Materials. All chemical substances and mixtures, including waste, which have toxic properties.

Toxic Pollutants. A specific list of 65 chemicals or classes of chemicals designated by EPA as toxic pollutants under the Clean Water Act.

Toxic Substances. Chemicals that are subject to the regulations issued under the Toxic Substances Control Act by EPA: this term is also used in a generic sense to mean toxic agents.

Toxic Wastes. Those hazardous wastes which are listed as toxic or meet the characteristics of toxicity as defined in RCRA regulations by EPA.

Treatment. Rendering a hazardous waste less hazardous, nonhazardous, or reduced in amount through processes which change its physical or chemical characteristics. Also used to refer to processes which make a waste easier to handle.

Trickling Filter. A bed of broken rock or other coarse aggregate onto which sewage or industrial waste is sprayed intermittently and allowed to trickle through, leaving organic matter on the surface of the rocks, where it is oxidized and removed by biological growths.

Turbidity. Any condition which causes a cloudy or hazy appearance in a naturally clear liquid, due to suspension of colloidal liquid droplets or fine solids.

Ultracentrifugation. Separation of liquid and solids by use of a centrifuge that develops centrifugal fields of more than 100,000 times gravity.

Ultrafiltration. Waste treatment employing membranes with extremely fine porosity.

Underground Injection. Use of specially designed wells to inject wastes into deep earth strata containing brine or brackish water; injection is almost always below the lowest drinking water aquifer.

Underground Storage Tank. All tanks containing regulated substances where the tank volume, including piping, is 10 percent or more beneath the surface of the ground.

Uniform Hazardous Waste Manifest. An EPA document; DOT authorizes its use as a shipping paper. A multipart form used to track offsite shipment of hazardous waste from generator to disposer.

Upper Explosive Limit. Maximum concentration of gas or vapor in air above which ignition does not occur.

Variance. Sanction granted by a governing body for delay or exception in the application of a given law, ordinance, or regulation.

Waste. Any discarded material.

Waste Stabilization Pond. Holding ponds in which no artificial aeration is introduced. Microbial degradation is aerobic near the surface; anaerobic digestion may occur at deeper levels.

SELECTED BIBLIOGRAPHY

The profit of books is according to the sensibility of the reader . . .
—Ralph Waldo Emerson

The following sources are those that have been the most valuable in the writing of this book. The list is by no means a complete record of all the sources that have been consulted, or all the sources that are available.

BOOKS

Arbuckle, J. G., G. W. Frick, R. M. Hall, Jr., M. L. Miller, T. F. P. Sullivan, and T. A. Vanderver, Jr. 1983. *Environmental Law Handbook*. 7th ed. Rockville, Md.: Government Institutes, Inc.

Bhatt, H. G., R. M. Syker, and T. L. Sweeney, eds. 1985. *Management of Toxic and Hazardous Wastes*. Chelsea, Mich.: Lewis Publishers.

Cahill, L. B., ed. 1984. *Environmental Audits*. 3d ed. Rockville, Md.: Government Institutes, Inc.

Cheremisinoff, P. N. 1986. *Hazardous Materials Book on Regulation*. Northbrook, Ill.: Pudvan Publishing Co.

Cheremisinoff, P. N., and K. A. Gigliello. 1983. *Leachate from Hazardous Wastes Sites*. Lancaster, Pa.: Technomic Publishing Co.

Congress of the United States Office of Technology Assessment. *Technologies and Management Strategies for Hazardous Waste Control*. Washington, D.C.: U.S. Government Printing Office.

Day, R. A., Jr., and A. L. Underwood. 1967. *Quantitative Analysis*. 2d ed. Englewood Cliffs, N.J.: Prentice-Hall.

Depol, D. R., and P. N. Cheremisinoff. 1984. *Emergency Response to Hazardous Materials Incidents*. Lancaster, Pa.: Technomic Publishing Co.

Duffy, J. L., ed. 1983. *Treatment, Recovery, and Disposal Processes for Radioactive Wastes*. Park Ridge, N.J.: Noyes Data Corporation.

Ecology and Environment, Inc., Buffalo, N.Y., and Whitman, Requardt and Associates, Baltimore, Md. 1985. *Toxic Substance Storage Tank Containment*. Park Ridge, N.J.: Noyes Publications.

Environmental Monitoring and Support Laboratory. 1979. *Handbook for Analytical Quality Control in Water and Wastewater Laboratories*. Cincinnati, Ohio: Environmental Protection Agency.

Friedlander, G., J. W. Kennedy, and J. M. Miller. 1964. *Nuclear and Radiochemistry*. New York: John Wiley and Sons.

Gimlin, H., ed. 1982. *Environmental Issues*. Washington, D.C.: Congressional Quarterly.

Glasstone, S., and W. H. Jordon. 1980. *Nuclear Power and Its Environmental Effects*. La Grange Park, Ill.: American Nuclear Society.

Harthill, M., ed. 1984. *Hazardous Waste Management—In Whose Backyard?* Boulder, Colo.: Westview Press.

Hazardous Waste Dialogue Group. 1983. *Siting Hazardous Waste Management Facilities*. White Plains, Md.: Automated Graphic Systems.

Honour, W. W. 1979. *Honours Energy and Environmental Handbook*. Westport, Conn.: Technomic Publishing Co.

Industrial Indemnity Company. 1980. *Safety Engineering Standards*. San Francisco, Calif.: Industrial Indemnity Co.

Institution of Civil Engineers. 1978. *Transport of Hazardous Materials*. London: Thomas Telford Ltd.

IU Conversion Systems, Inc. 1979. *A Glossary of Terms, Expressions and Acronyms as Used in the Solid and Liquid Waste Management Field*. Horsham, Pa.: IU Conversion Systems.

Lapedes, D. N., ed. 1974. *Dictionary of Scientific and Technical Terms*. New York: McGraw-Hill.

Lester, J. P., and A. O'M. Bowman, eds. 1983. *The Politics of Hazardous Waste Management*. Durham, N.C.: Duke University Press.

Lipschutz, R. D. 1980. *Radioactive Waste*. Cambridge, Mass.: Ballinger Publishing Co.

Lowry, G. G., and R. C. Lowry. 1985. *Handbook of Hazard Communication and OSHA Requirements*. Chelsea, Mich.: Lewis Publishers.

Majumdar, S. K., and E. W. Miller, eds. 1985. *Management of Radioactive Materials and Wastes: Issues and Progress*. Phillipsburg, N.J.: Typehouse of Easten.

Meyer, E. 1977. *Chemistry of Hazardous Materials*. Englewood Cliffs, N.J.: Prentice-Hall.

National Safety Council. 1985. *The Hazard Communication Standard*. Chicago: National Safety Council.

Nelkin, D., and M. S. Brown. 1984. *Workers at Risk*. Chicago: University of Chicago Press.

Office of Pesticides and Toxic Substances. 1985. *Guidance for Controlling Asbestos-Containing Materials in Buildings*. Washington, D.C.: Environmental Protection Agency.

Office of Public Affairs. 1986. *Asbestos Fact Book*. 3d rev. Washington, D.C.: Environmental Protection Agency.

Office of Solid Waste. 1985. *Asbestos Waste Management Guidance*. Washington, D.C.: Environmental Protection Agency.

Office of Technology Assessment. 1983. *Technologies and Management Strategies for Hazardous Waste Control*. Washington, D.C.: U.S. Government Printing Office.

Peirce, J. J., and P. A. Vesilind, eds. *Hazardous Waste Management*. Ann Arbor, Mich.: Science Publishers.

Quarles, J. 1982. *Federal Regulation of Hazardous Wastes*. Washington, D.C.: Environmental Law Institute.

Shields, E. J. 1985. *Pollution Control Engineer's Handbook*. Northbrook, Ill.: Pudvan Publishing Co.

Smith, A. J., Jr. 1981. *Managing Hazardous Substances Accidents*. New York: McGraw-Hill.

Toxic Waste Assessment Group, Governor's Office of Appropriate Technology. 1981. *Alternatives to the Land Disposal of Hazardous Wastes*.

Truitt, T. H., D. R. Berz, D. B. Weinberg, J. B. Malloy, G. L. Price, and B. T. Florence. 1981. *Environmental Audit Handbook*. New York: Executive Enterprises Publications Co.

U.S. Department of Transportation. 1984. *1984 Emergency Response Guidebook*. Chicago: Labelmaster, Div. Modular Products Corporation.

U.S. Environmental Protection Agency. 1985. *The Layman's Guide to the Toxic Substances Control Act*. Washington, D.C.: Office of Toxic Substances.

U.S. Forest Service. 1983. *Transportation of Hazardous Materials*.

Worobec, M. D. 1984. *Toxic Substances Controls Primer*. Washington, D.C.: Bureau of National Affairs.

ARTICLES

Abelson, P. H. 1984. "Environmental Risk Management." *Science* (November): 3.

Adessa, M. 1987. "Waves in the Water Works." *Civil Engineer* (March): 48-50.

Alm, A. L. 1985. "Environmental Management Counts." *The Environmental Forum* (October): 32-35.

Associated Press. 1987. "EPA Won't Ban Pesticides While Testing Goes On." *The Gainesville Sun* (28 January).

Bayer, J. E. 1984. "New Routes for Hazardous Waste Management." *Pollution Engineering* (September): 25-27.

Bernstein, C. S. 1985. "Hammering Out a New RCRA." *Civil Engineering* (April): 57-60.

Berg, G. 1987. "Disaster Preparedness: A Way to Get Ready." *Environment* (April): 27-30.

Borner, A. J. 1985. "Moving Towards a Workable Contingency Plan." *The Environmental Manager's Compliance Advisor* (1 April): 4-6.

Bower, J. 1987. "Management of Environmental Information." *Pollution Engineering* (January): 52-55.

Bradford, H. 1987. "Firms Must Disclose Hazards." *Engineering News Record* (27 August): 10.

"Brakes a Hitch to Complete Asbestos Ban?" 1987. *Engineering Times* (June).

Brown, J. 1986. "1986 Environmental Software Review." *Pollution Engineering* (January): 18-22.

Bula, F. 1987. "OSHA—A Sleeper." *Environmental & Waste Management World* (June): 9-11.

Bureau of Law & Business, Inc. 1985. "Dictate the Tone of the Inspection: Have an Inspection Protocol." *The Environmental Manager's Compliance Advisor* (May): 6.

"Burning Toxics." 1987. *Engineering News Record* (16 July): 22.

Camouis, G. 1985. "Toxic Materials Risk Assessment: A Practical Guide." *Pollution Engineering* (August): 50-57.

Cheremisinoff, P. N. 1986. "Special Report: Treatment of Hazardous Wastes." *Pollution Engineering* (November): 29-44.

______. 1987. "Update: Hazardous Waste Treatment." *Pollution Engineering* (February): 42-48.

Cheremisinoff, P. N., J. G. Casana, and H. W. Pritchard. 1986. "Update on Underground Tanks." *Pollution Engineering* (August): 12-25.

Cheremisinoff, P. N., and J. T. Eyck. 1987. "Environmental Auditing: A Basic Guide." *Pollution Engineering* (April): 72-75.

"Community Right-to-Know Rules and List of Extremely Hazardous Substances Finalized." 1987. *The Hazardous Waste Consultant* (July/August): 29-31.

Corbitt, R. A., D. C. Garrett III, R. F. Kohm, D. L. Patrick, and D. R. Perander. 1985. "Auditors Eye the Environment." *Civil Engineering* (April): 39-41.

Cullinane, J., Jr., and R. A. Shafer. 1985. "Waste Cleanup: Lessons Learned." *Civil Engineering* (June): 41-43.

Culp, F. 1987. "Big Changes Ahead for Drinking Water Industry." *Water Engineering & Management* (March): 24-29.

Dahlsten, D. 1983. "Pesticides in an Era of Integrated Pest Management." *Environment* (December): 45-54.

Darst, G. 1987. "After 17 Years, Clean-Air Standards Remain Beyond Reach of Large Cities." *The Gainesville Sun* (3 May).

Davis, J. 1987. "Nuclear Waste." *The Gainesville Sun* (22 March).

Dombrowski, C. H. 1987. "Ground in Disposal Options." *World Wastes* (August): 43-48.

______. 1986. "The EPA Updates Rules for Hazwaste Sites." *World Wastes* (July): 79-80.

______. 1986. "Superfund Passes Despite Tax Feud." *World Wastes* (December): 15.

Doria, J. 1979. "Radiation Regulation." *Environment* (December): 25-31.

"EPA Adds to Superfund List." 1987. *Engineering News Record* (30 July): 13.

"EPA Modifies Clean Air Regs." 1987. *Engineering News Record* (11 June): 15.

"EPA Releases Two Lists of Hazardous Chemicals." 1987. *Engineering News Record* (30 April): 14-15.

"EPA Wants Leak-detection System for New Hazardous-Waste Dumps." 1987. *Engineering News Record* (28 May): 7.

Frankel, M. L. 1983. "Report on Reports." *Environment* (July/August): 25-27.

Ghassemi, M. 1987. "In Situ Technologies for Site Decontamination." *Environmental & Waste Management World* (July): 6-8.

Gray, R. 1987. "Washington News." *Water Engineering & Management* (8 April): 5.

Guida, J. F. 1983. "Corporate Decision Making to Meet EPA Hazardous Waste Requirements." *Professional Engineer* (Spring): 15-19.

Hartley, J. 1986. "New Tools for an Old but Escalating Problem." *Consulting Engineer* (May): 42-48.

"Hazardous-Waste Treatment Undergoes Dramatic Improvement." 1986. *Chemical Engineering* (10 November): 67.

Heil, J., and J. Van Blaroom. 1986. "Superfund: The Search for Consistency." *Environment* (April): 6-9.

Hersch, P. 1985. "Management Can Take Inventory of Itself." *Water Engineering & Management* (July): 16.

Hickerson, K. W. 1986. "Clean Sites, Inc., Sets Impressive Record for Cleanup Settlements." *Engineering Times* (June): 8.

Hirschhorn, J. S., and K. V. Oldenburg. 1987. "Taking Seriously the Reduction of Hazardous Waste." *Pollution Engineering* (January): 50-51.

Hoffmann, P., ed. 1987. "Washington Observer." *Engineering News Record* (23 April): 7.

Hope, S. J. 1986. "Constructing a Better Environment." *Civil Engineering* (October): 66-68.

Hope, W. C. 1983. "Information Management." *Pollution Engineering* (January): 22-23.

Hunsaker, C., and J. Kelly. 1987. "A Watery Grave." *Civil Engineering* (March): 52-55.

Karagans, J. 1987. "A Purity Crusade Takes on the Drinking Water Industry." *Water Engineering & Management* (March): 27-29.

Klein, S. N. 1983. "How to Survive an EPA Inspection." *Nation's Business* (August): 28-29.

McCoy and Associates. 1984. "EPA Finalizes Uniform Waste Manifest Requirements." *The Hazardous Waste Consultant* (May/June): 8-9.

______. 1984. "Regulatory Issues." *The Hazardous Waste Consultant* (November/December): 1-6.

______. 1984. "Special Features." *The Hazardous Waste Consultant* (November/December): 1-18.

Michaels, M. 1986. "Cleaning Up the Image of Wastes." *World Wastes* (April): 93-96.

Moses, G. 1987. "EPA Official Praises Groundwater Program." *The Gainesville Sun* (5 April).

"NBS Develops Smoke Toxicity Measures." 1987. *Engineering Times* (February): 13.

Newton, J. 1986. "Community Awareness and Emergency Response." *Pollution Engineering* (July): 20-27.

"OSHA Wants to Negotiate as It Sets New Standards." 1987. *Engineering News Record* (23 July): 7.

Ouellette, R. P., and P. N. Cheremisinoff. 1987. "Asbestos Hazard Management." *Pollution Engineering* (March): 36-43.

Parker, F. J. 1986. "The Insurance Crisis and Environmental Protection." *Environment* (April): 14-18.

Rich, G. A. 1987. "Crisis Management: Preparation Checklist." *Pollution Engineering* (February): 62-64.

______. 1986. "Weighing the Factors about Treating Hazardous Wastes." *Pollution Engineering* (July): 26-27.

"Risky Waters." 1987. *Civil Engineering* (April): 68-71.

Salim, R. 1987. "Floodgates Open for Clean Water Act Drown Second Presidential Veto." *Engineering Times* (March): 1-3.

"Separate But Equal." 1987. *Engineering News Record* (August): 66.

Shabecoff, P. 1987. "EPA Study Cites Pesticide Cancer Risk." *The Gainesville Sun* (15 April).

Sittenfield, M. 1983. "A Hazardous Waste Audit." *Professional Engineer* (Spring): 20-23.

Stack, V. 1987. "Waste Water Technology Options Today." *Water Engineering & Management* (June): 26-27.

Steinway, D. M. 1987. "The New Superfund: Summary of Its Reauthorization." *Pollution Engineering* (January): 46-49.

______. 1987. "Regulation of Asbestos in Public and Commercial Buildings: Impending Federal/State Governmental Activity." *Pollution Engineering* (July): 16.

Thompson, J. C. 1986. "The Safe Drinking Water Act." *Public Works* (September): 123-125.

______. 1986. "Updating the Safe Drinking Water Regulations." *Water Engineering & Management* (August): 21-24.

"Tire Change for Transport of Hazardous Materials." 1987. *Engineering Times* (June).

Tschinkel, V. J. 1986. "The Transition Toward Long-Term Management." *Environment* (April): 19-26.

Tusa, W. 1986. "Risk Assessment: Engineering Tool." *Civil Engineering* (February): 59-61.

"Underground Tank Rules Are on Verge of Release." 1987. *Engineering News Record* (March): 14.

INDEX

Absorption, 142
Accuracy, 17-18
Activated sludge, 145
Administrative Procedure Act, 20-22
Adsorption, 142
Advance Notice of Proposed Rulemaking, 20
Aerated lagoons, 145
Agreement States, 119
Air Quality Control Regions, 54
Air stripping, 142
Aldrin, 34
Anaerobic digestors, 145
Analytical procedures, 14-17
Arbitration, 174
Asbestos, 37, 90, 109. *See also* Asbestos Control Regulations
Asbestos abatement activities, 110-111, 258-260
Asbestos Control Regulations: background, 109-110; disposal regulations, 114; enforcement, 114; potential future developments, 114-115; related to asbestos in buildings, 110-114, 255; relationship to other regulations, 37, 114
Asbestos Hazard Emergency Response Act of 1986, 109, 111, 113
Asbestos-in-Schools Rule, 111, 113
Asbestos School Hazard Abatement Act, 109, 113
BLEVE (Boiling Liquid Expanding Vapor Explosion), 8-10

CAA. *See* Clean Air Act
Cancer policy. *See* Occupational Safety and Health Act
Centrifugation, 141
CERCLA. *See* Comprehensive Environmental Response, Compensation, and Liability Act
CFCs (Chlorofluorocarbons), 90
CFR. *See* Code of Federal Regulations
Chemicals-in-Commerce Information System, 89
Chemical Substances Information Network, 89
Chlordane, 34
Clean Air Act (CAA): background, 49-50; enforcement, 58, 61; goals, 50; imminent hazard provision, 58; motor vehicle regulations, 57; potential future developments, 62; relationship to other regulations, 34, 46, 58-59, 69, 91, 109-114; standards in, 50-60
Clean Sites Inc., 131, 133
Clean Water Act (CWA): background, 49, 62, 63; disposal of sewage sludge, 67; enforcement, 67; imminent hazard provision, 67; pollutants regulated, 2, 63, 182-184;

potential future developments, 69; pretreatment standards, 64; relationship to other regulations, 34, 59, 67-69, 86, 91, 107, 114, 151
Coalitions, 130, 131, 133
Codes of Federal Regulations (CFR), 3, 20-22
Comprehensive Environmental Response, Compensation, and Liability Act (CERCLA): background, 77, 101; enforcement, 107; funding of, 105; liability under, 105; notification requirements, 105-106; potential future developments, 107-108; relationship to other regulations, 42, 69, 76, 86, 99, 100, 107; response actions, 102-105
Computerized information management systems, 130, 131-132
Construction, management of hazardous materials during, 151
Contingency planning, 157-166
Corrosive materials, 7
Cost-benefit analysis, 127
Council on Environmental Quality, 1, 44-47, 89
Cryogens, 10
CWA. *See* Clean Water Act

Dechlorination, 143
Delisting, 136, 137, 140
Dioxins, 90
Disposal: landfills, 148, 150; ocean dumping, 150-151; underground injection wells, 70, 73-74, 150
Distillation, 141-142
Dredge and fill activities, 64-65

Effluent guidelines, 65, 66
Effluent limitations, 65, 66
Emergency information systems, 162, 164
Emergency Planning and Community Right-to-Know Act, 106
Emergency prevention, 170
Emergency response: contingency planning, 157-166; evacuation, 164; fire protection, 166; identification procedures, 159-162; information systems, 162, 164; notification procedures, 162-164; protective devices, 166-170; spill containment, 165-166
Enforcement, 22. *See also enforcement sections under individual environmental laws by name*
Environmental audits, 130, 172-173
Environmental Impact Statement, 44-47
Environmental laws, 2, 19-27. *See also individual environmental laws by name*
Environmental Protection Agency. *See* EPA
EPA (Environmental Protection Agency), 3, 31, 36, 44. *See also enforcement sections under individual environmental laws*
Error, 15-17
Evacuation, 164
Evaporation, 141
Exchange, 136, 137, 139-140
Explosives, 8, 83
Exposure, 11

Federal Emergency Management Agency (FEMA), 157, 158, 162
Federal Environmental Pesticide Control Act, 31
Federal Insecticide, Fungicide, and Rodenticide Act (FIFRA): background, 29-31; control mechanisms, 31-33; enforcement, 33-34; potential future developments, 34; relationship to other regulations, 34, 41, 59, 92
Federal Register, 2, 3, 20-22
Federal Water Pollution Control Act, 63
FEMA. *See* Federal Emergency Management Agency
FIFRA. *See* Federal Insecticide, Fungicide, and Rodenticide Act
Filtration, 141
Fire danger (fire hazard), 6-10
Fire protection, 166
Formaldehyde, 43
Freedom of Information Act, 20

Groundwater monitoring, 98, 150
Groundwater protection. *See* Safe Drinking Water Act

Hazard, 1
Hazard Communication Standard, 37, 38-42
Hazard Ranking System, 103, 104
Hazardous, 5
Hazardous and Solid Waste Amendments of 1984, 98-99
Hazardous gases, 11
Hazardous materials, 1
Hazardous Materials Transportation Act (HMTA): allowable packages and containers, 79-80; background, 77-78; classification of materials, 81, 193-196; enforcement, 84, 86; labeling, 81, 83-84; marking, 81-83; placarding, 81, 84; relationship to other regulations, 42, 86, 100, 107, 114; shipping paper requirements, 84-85
Hazardous properties, 5-6
Hazardous substances, 1
Hazardous waste, 2. *See also* Resource Conservation and Recovery Act
Hazard Ranking System, 103, 104
Health standards. *See* Occupational Safety and Health Act
Heptachlor, 34
HMTA. *See* Hazardous Materials Transportation Act
Household wastes, 153-156

Imminent hazard provisions, 22-23. *See also individual environmental laws by name*
Incineration, 146-148
Inspections, 173-174
Insurance, 127, 130, 133
Ion exchange, 143

Judicial review, 22

Knowing endangerment, 100

Land farming, 145
Landfills, 148, 150
Law: definition of, 19; state and local, 26. *See also individual environmental laws by name*
Leaking Underground Storage Tanks, 98
Lethal concentration, 11
Lethal dose, 11
Low-Level Radioactive Waste Policy Act of 1980, 118, 121-122

Management, 125. *See also* Program Management, Risk Management, Substance Management
Management inventories, 130, 131
Material Safety Data Sheet, 39-40
Maximum Contaminant Level Goals, 72
Maximum Containment Levels, 71-72
Membrane separation processes, 141
Midnight dumping, 150
Minimization, 135, 136, 137, 139
Monitoring, 129, 148
Monitoring wells, 98

NAAQS. *See* National Ambient Air Quality Standards
National Ambient Air Quality Standards (NAAQS), 50, 51-54, 57-58, 60
National Contingency Plan (NCP), 65, 69, 101, 104, 107, 158
National Emission Standards for Hazardous Air Pollutants (NESHAPs), 50, 54-56, 60
National Environmental Policy Act (NEPA): applicability, 45; background, 29, 43-44; enforcement, 46; potential future developments, 46-47; relationship to other regulations, 46, 69
National Fire Protection Association 704 Marking System, 160-161
National Institute of Occupational Safety and Health (NIOSH), 35, 37, 38
National Pollution Discharge Elimination System, 64-66
National Priorities List, 102, 103, 104, 222-240
NCP. *See* National Contingency Plan

NEPA. *See* National Environmental Protection Act
NESHAPs. *See* National Emission Standards for Hazardous Air Pollutants
Neutralization, 142
New Source Performance Standards, 57-60
NIOSH. *See* National Institute of Occupational Safety and Health
NRC. *See* Nuclear Regulatory Commission
Nuclear Regulatory Commission (NRC), 117-122
Nuclear Waste Policy Act of 1982, 118, 121

Occupational Safety and Health Act (OSH Act): background, 29, 35-36; cancer policy, 37, 38; enforcement, 41-42; Hazard Communication Standard, 37, 38-42; Hazardous Waste Operations and Emergency Response Rules, 37, 39; health standards, 37-38; potential future developments, 43; relationship to other regulations, 34, 91, 107, 109-114
Occupational Safety and Health Administration (OSHA), 37, 41-42, 117, 122
Ocean dumping, 150-151
Organic compounds, 7
OSHA. *See* Occupational Safety and Health Administration
OSH Act. *See* Occupational Safety and Health Act
Oxidation/reduction processes, 142
Oxidizing agents, 7, 9

PCBs (Polychlorinated biphenyls), 90, 137, 143, 148
Pesticides, 31. *See also* Federal Insecticide, Fungicide, and Rodenticide Act
Point source discharges, 49, 62, 77
Polychlorinated Biphenyls. *See* PCBs
Polymers, 7
Precipitation, 142
Precision, 17-18
Premanufacturing Notice, 88
Pretreatment standards, 64
Prevention of Significant Deterioration, 55, 58, 61
Primary drinking water regulations, 70-72
Primary organic hazardous constituent, 146
Priority pollutants, 63, 182-184
Program Management: cost effectiveness of, 127; future needs, 133-134; goals, 125-127; implementation, 128-130; innovative approaches, 130-133
Protective devices: decontamination of, 170; overprotection, 170; protective clothing, 166, 168-170; respiratory protective equipment, 170
Public relations, 129-130
Pyrolysis, 148

Quality assurance, 15-17
Quality control, 15-17

Ratioactive materials: exposure limits, 13, 14; genetic effects, 13; somatic effects, 12
Radioactive materials: exposure limits, tions: background, 117-118; enforcement, 122; potential future developments, 122-123; relationship to other regulations, 59, 76, 122; transportation regulations, 120-121; usage regulations, 119-120; waste management regulations, 121-122
Radioactivity, 6
RCRA. *See* Resources Conservation and Recovery Act
Rebuttal Presumption Against Registration, 32-33
Recommended Maximum Contaminant Levels, 71-72
Recovery, 136, 137, 139-140
Recycling, 136, 137, 139-140
Reduction, 135, 136, 137, 139
Regulations: changes in, 26; chronological order of, 22-23; definition of, 19; development of, 20-22; enforce-

ment of, 22; judicial review of, 22; obtaining and updating of, 3, 19-20; prospective focus of, 101. *See also individual environmental laws by name*
Regulatory compliance: definition of, 171-172; documents used to determine, 19; inspections for, 173-174; profile, 173
Rem, 13
Reportable quantities, 65, 82, 86, 106, 185-188
Resource Conservation and Recovery Act (RCRA): background, 77, 92-93; enforcement, 92, 99-100; generator requirements, 96-97; potential future developments, 100-101; relationship to other regulations, 34, 42, 46, 58-59, 76, 82, 86, 92, 100, 107; transporter requirements, 96-97; treatment, storage, and/or disposal facility requirements, 92, 96-98; underground storage tank regulation, 98-99; waste classification, 2, 92, 93-96, 197-221
Reuse, 136, 137, 139-140
Right-to-Know laws, 39, 42. *See also* Hazard Communication Standard
Risk assessment, 128-129
Risk Management, 127-129

Safe Drinking Water Act: background, 49, 69-70; enforcement, 74-75; imminent hazard provision, 74; potential future developments, 72, 76, 191-192; primary drinking water regulations, 70-72; relationship to other regulations, 76, 100, 107; sole source aquifer regulation, 74; Underground Injection Control, 70, 73-74
Safety, 125-126
Separation processes, 141
Shipping papers, 84-85
Significant New Use Rule, 88, 92
Smoke, 11
Sole source aquifers, 74
Solidification, 143-144
Solvent extraction, 142
Specialized microbial strains, 145-146
Spill containment, 165-166
Spill Prevention Control and Countermeasure plans, 64
Stabilization. *See* Solidification
State Implementation Plans, 58
Statute, 19
Steam stripping, 142
Storage, 135, 136, 137-138, 261-266
Stress: definition of, 6; effects, 13-14
Substance Management: background, 135-137; delisting, 136, 137, 140; disposal, 136, 137, 148, 150-151; during construction, 151; facilities, 151-152; household wastes, 153-156; innovative technologies, 148-149; minimization/reduction, 135, 136, 137, 139; recycling/reuse/recovery/exchange, 136, 137, 139-140; storage, 135, 136, 137-138; treatment, 136, 137, 140
Suggested No Adverse Response Levels, 72
Superfund. *See* Comprehensive Environmental Response, Compensation, and Liability Act
Superfund Amendments and Reauthorization Act, 106-107, 241-254; relationship to other regulations, 34, 92

Threshold Limit Value, 11
Threshold Planning Quantities, 106, 241-252
Toxic agents. *See* Toxic substances
Toxic chemicals. *See* Toxic substances
Toxicity, 2, 10-11
Toxic Materials, 2
Toxic pollutants, 2, 63, 65, 182-184
Toxic substances, 2
Toxic Substances Control Act (TSCA): background, 77, 87; chemicals regulated, 2, 87; control measures, 88, 89-90; enforcement, 91; imminent hazard provision, 90-91; potential future developments, 92; relationship to other regulations, 41, 91-92, 109-114
Toxic wastes, 2
Treatment: biological, 144-146; chemi-

cal, 142-144; incineration, 146-148; innovative, 148-149; physical, 140-142
Trickling filters, 145
TSCA. *See* Toxic Substances and Control Act

Underground Injection Control, 70, 73-74
Underground injection wells, 72-74, 150
Underground storage tanks, 98-99
UN hazard class numbers, 84, 160
Uniform Hazardous Waste Manifest, 84-85, 96
United States Code, 3, 19

Waste exchanges, 139-140
Waste stabilization ponds, 145
Water Quality Act of 1987, 67, 189, 191-192
Water quality criteria, 65, 66, 70
Water quality standards, 65, 66
Water reactive materials, 7, 8

About the Author

AILEEN SCHUMACHER is President of Blum, Schumacher, & Associates, Inc., which provides environmental consulting engineering services in addition to civil and structural engineering and inspection services.